CAMBRIDGE LIBRARY COLLECTION

Books of enduring scholarly value

Earth Sciences

In the nineteenth century, geology emerged as a distinct academic discipline. It pointed the way towards the theory of evolution, as scientists including Gideon Mantell, Adam Sedgwick, Charles Lyell and Roderick Murchison began to use the evidence of minerals, rock formations and fossils to demonstrate that the earth was older by millions of years than the conventional, Bible-based wisdom had supposed. They argued convincingly that the climate, flora and fauna of the distant past could be deduced from geological evidence. Volcanic activity, the formation of mountains, and the action of glaciers and rivers, tides and ocean currents also became better understood. This series includes landmark publications by pioneers of the modern earth sciences, who advanced the scientific understanding of our planet and the processes by which it is constantly re-shaped.

A Selection of the Geological Memoirs Contained in the Annales des Mines

Sir Henry Thomas de la Beche (1796–1855) was a talented and influential geologist. A friend of Mary Anning, he produced the famous lithograph *Duria antiquior* (1830), the first reconstruction of a scene from an ancient world, to support her work. He promoted government involvement in geology and became the founding Director of the British Geological Survey, which was officially recognised in 1835. Inspired by his work in Cornwall, he later founded the Royal School of Mines and the Museum of Practical Geology. Among his published works was a *Manual of Geology* (1831), which went through three English editions and was published in France, Germany and America. This 1824 collection of translations includes studies on sites across Europe and notes on the production of an early geological map of France. He also provides a table of equivalent formations and a translation of Brongniart's *Classification of the Mixed Rocks.*

Cambridge University Press has long been a pioneer in the reissuing of out-of-print titles from its own backlist, producing digital reprints of books that are still sought after by scholars and students but could not be reprinted economically using traditional technology. The Cambridge Library Collection extends this activity to a wider range of books which are still of importance to researchers and professionals, either for the source material they contain, or as landmarks in the history of their academic discipline.

Drawing from the world-renowned collections in the Cambridge University Library and other partner libraries, and guided by the advice of experts in each subject area, Cambridge University Press is using state-of-the-art scanning machines in its own Printing House to capture the content of each book selected for inclusion. The files are processed to give a consistently clear, crisp image, and the books finished to the high quality standard for which the Press is recognised around the world. The latest print-on-demand technology ensures that the books will remain available indefinitely, and that orders for single or multiple copies can quickly be supplied.

The Cambridge Library Collection brings back to life books of enduring scholarly value (including out-of-copyright works originally issued by other publishers) across a wide range of disciplines in the humanities and social sciences and in science and technology.

A Selection of the Geological Memoirs Contained in the Annales des Mines

Edited and translated by
Henry T. De La Beche

CAMBRIDGE UNIVERSITY PRESS

Cambridge, New York, Melbourne, Madrid, Cape Town,
Singapore, São Paolo, Delhi, Mexico City

Published in the United States of America by Cambridge University Press, New York

www.cambridge.org
Information on this title: www.cambridge.org/9781108048408

This edition first published 1824
This digitally printed version 2012

ISBN 978-1-108-04840-8 Paperback

The original edition of this book contains a number of colour plates, which have been reproduced in black and white. Colour versions of these images can be found online at www.cambridge.org/9781108048408

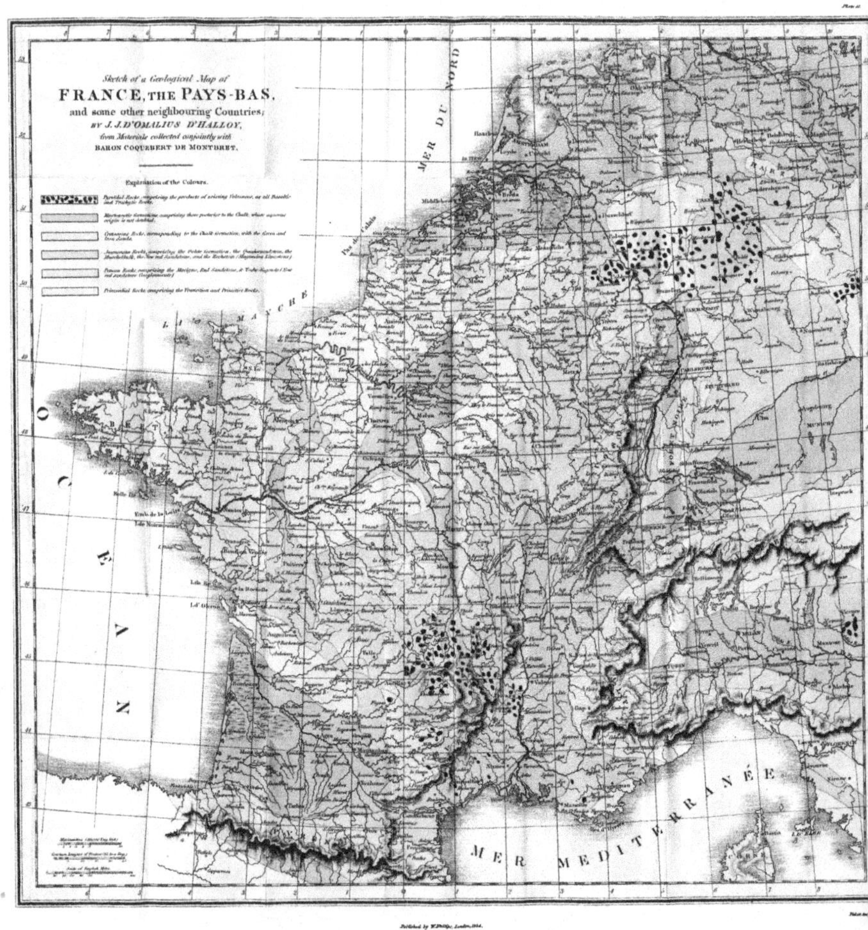

The material originally positioned here is too large for reproduction in this reissue. A PDF can be downloaded from the web address given on page iv of this book, by clicking on 'Resources Available'.

A SELECTION OF THE GEOLOGICAL MEMOIRS CONTAINED IN THE ANNALES DES MINES,

TOGETHER WITH A SYNOPTICAL TABLE OF

EQUIVALENT FORMATIONS,

AND

M. BRONGNIART'S TABLE OF

THE CLASSIFICATION OF MIXED ROCKS.

TRANSLATED, WITH NOTES,

BY H. T. DE LA BECHE, ESQ. F.R.S. F.L.S.

MEM. GEOL. SOC. MEM. PHYS. AND NAT. HIST. SOC. OF GENEVA, &c.

LONDON:

PRINTED AND SOLD BY WILLIAM PHILLIPS,

GEORGE YARD, LOMBARD STREET.

1824.

PREFACE.

I HAVE endeavoured in the following pages to collect much valuable Geological information, which was scattered through several volumes of the Annales des Mines, a work edited by the General Council of Mines at Paris; and it is hoped that by thus presenting it in a portable form, it may be found useful to English Geologists, by enabling them easily to compare some of our own rocks with similar formations described in the above work; a correct knowledge of the changes which take place in many rocks, particularly those of the secondary class, in their mineralogical structure, &c. at different distances, being I conceive not one of the least important branches of geological inquiry.

In order to avoid a repetition of synonymous terms, I have prefixed a Synoptical Table of Equivalent Formations; the arrangement adopted has been simply that of convenience, and I wish it to be clearly understood that it is no attempt at geological divisions: from the alluvium to the greywacke and its associated beds, the rocks are placed in the order of superposition; beneath these the other formations are sometimes so mixed, that it would be difficult to assign them a regular mode of occurrence; granite and gneiss have however, according to the usual practice, been placed at the bottom of the list; the rocks usually termed trap rocks, have been thrown together for the convenience of more easy reference. It is not always very clear whether our lower chalk without flints should be referred to the lower portion of the upper white chalk of the French, or to their *craie tufau.* I have inserted the muschelkalk and quadersandstein of Germany as separate formations, in order to shew the opinions at present entertained by some continental geologists on the subject, who consider the muschelkalk as distinct from our lias; it has on the other hand been considered by some geologists that the muschelkalk is a modification of our lias, and that the quadersandstein* is the sand of the inferior oolite; conceiving it to be of some importance to determine if we are or are not to add two new formations to our secondary rocks, I have, in an Appendix, given the descriptions of these rocks by Messrs. Humboldt and Boué, for the sake of comparison. With respect also to the identity of the new red conglomerate with the rothe todte liegende of Germany, it may perhaps be right to mention, as discussions have lately taken place on this subject

* This only refers to those localities where the sandstone thus termed is distinctly interposed between the muschelkalk and Jura limestone, for it is impossible to resist the suspicion, that more than one sandstone formation has been confounded under this name in different districts.

between the Rev. W. D. Conybeare and Mr. Weaver, that the conglomerate usually termed new red conglomerate, in the neighbourhood of Exeter and Teignmouth, seems closely to resemble the rothe todte liegende, as has been already stated by Prof. Buckland; the magnesian limestone is unfortunately wanting in that country, or at least has not been described, though traces of it are mentioned by Mr. Conybeare (Outlines of the Geology of England and Wales, p. 308) at Sampford Peverell in Devonshire, for beneath that rock the German rothe todte liegende is always described as occurring.

In addition to the Synoptical Table above mentioned, will be found a translation of M. Brongniart's table of the Classification of the Mixed Rocks, inserted in the Journal des Mines, which will be found useful, by shewing the composition of rocks, bearing names not in common use amongst English geologists.

Some of the Memoirs will be found shortened in the following translations, but it is hoped that nothing material has been omitted.

H. T. De la Beche.

CONTENTS.

CONTENTS.

TABLE OF THE CLASSIFICATION

OF THE

MIXED ROCKS.

BY ALEXANDRE BRONGNIART.

(Extracted from the JOURNAL DES MINES, No. 199, July, 1813.)

CLASS I. CRYSTALLIZED ISOMEROUS (ISOMERES) ROCKS.

Character. The parts united by crystalline aggregation, without either an essential predominant base or portion, or a perceptible homogeneous cement.

Genus 1. FELSPATHIC ROCKS.

Char. Felspar is the essential constituent part.

Species 1. GRANITE.

Essentially composed of lamellar felspar, quartz, and mica, nearly equally disseminated.

Principal varieties.

COMMON GRANITE..... Felspar, quartz, and mica, equally disseminated.

PORPHYRITIC GRANITE. Crystals of felspar in small grained granite.

Species 2. PROTOGINE. (*Jurine.*)

Essentially composed of felspar, quartz, and steatite, talc, or chlorite, entirely or almost entirely replacing the mica.

Observation. The decomposing (altérées) syenites are easily confounded with this rock.

(The Pomenaz, valley of Servos,—the Talèfre,—the Gorge of Mallavale in Oisans,—the Sonnenberg in the Hartz,—Niolo in Corsica.)

Species 3. PEGMATITE. (*Haüy.*) *Graphic granite*, &c.

Essentially composed of lamellar felspar and quartz.

Obs. All the fine kaolins are derived from this rock.

(St. Yriex, near Limoges,—Geyer,—Cambo, near Bayonne,—Longcrup, near Bagnères)

Species 4. MIMOSE. (*A species establihed by M. Haüy.*)

Essentially composed of augite and lamellar felspar.

(Summit of the Meissner.)

Genus 2. HORNBLENDE ROCKS.

Char. Hornblende is the essentially constituent part.

Species 1. SYENITE. (*Werner.*)

Essentially composed of lamellar felspar, hornblende, and quartz. The felspar is often predominant.

Principal varieties.

GRANITIC S. ....Felspar and lamellar hornblende with a little mica.
(Upper Egypt,—Plauen in Saxony,—the Rehberg in the Hartz.)

SCHISTOSE S. ....Lamellar felspar and hornblende with a slaty structure.

PORPHYRITIC S. ......Large felspar crystals in a small grained syenite. (Altenberg in Saxony.)

ZIRCON S.............Felspar, lamellar hornblende, and zircon. (Fridrichwern in Norway.)

Species 2. DIABASE. (Grünstein, *Werner.*)

Essentially composed of hornblende and compact felspar nearly equally disseminated.

Principal varieties.

GRANITIC D. ....The structure granular.
(The diabase of the ancient Egyptian monuments; it contains black mica. La Perque, opposite Coutance, &c.)

SCHISTOSE D. ....The structure fissile, rayed, or zoned.
(Charbiac, near Saint-Flour. The Chalanches in Oisans,—the Schneeberg.)

PORPHYRITIC D. .....Crystals of compact felspar disseminated in a fine grained greenstone.

ORBICULAR D. ....Spheres with concentric zones of hornblende and compact felspar in a diabase of moderately sized grains.
(Orbicular granite of Corsica.)

Species 3. HÉMITHRENE.

Essentially composed of hornblende and limestone.

(The rock named primitive greenstone with limestone of Andreasberg, in the Hartz,—of Smalzgrube in Saxony. The rock named limestone of Maresberg in Saxony.)

CLASS II. THE CRYSTALLIZED ANISOMEROUS (ANISOMERES) ROCKS.

Char. Wholly or in part formed by confused crystallization; one predominant part serving as a base, paste or cement to the others, and contemporaneous, or anterior to thé parts it contains.

Genus 1. WITH A BASE OF CRYSTALLIZED QUARTZ.

Species 1. HYALOMICTE. (*Graisen.*)

Essentially composed of crystallized quartz, with mica disseminated but not continuous. The structure granular.

(Altemberg, with tin,—Vaulry, near Limoges, with wolfram.)

Genus 2. WITH A BASE OF MICA.

Species 1. GNEISS.

Essentially composed of an abundance of mica in plates, and of lamellar or granular felspar,—the structure laminated.

Obser. As there are rocks named gneiss by all geologists, which do not contain quartz, this mineral cannot be termed an essential constituent part.

Principal varieties.

COMMON GN..........	Little or no quartz.
QUARTZOSE GN.......	Quartz abundant. (Todstein in Saxony,—Huttenberg, Hartz,)
TALCOSE GN..........	Granular felspar, shining and talcose mica. (Saint Bel, near Lyon.)
PORPHYRITIC GN......	Crystals of felspar disseminated in gneiss. (Kringeln, in Norway (*von Buch.*) Cevin in the Tarentaise.)

Species 2. MICASCHISTE. (Mica Slate.)

Essentially composed of an abundance of continuous mica and quartz. The structure slaty.

Principal varieties.

QUARTZOSE M........	Very abundant quartz and mica, nearly alone, and alternating in undulating laminæ.
GRANITIC M.	Garnets nearly equally disseminated.
FELSPATHIC M........	Grains of felspar nearly equally disseminated. (Herold, near Ehrenfriedersdorf in Saxony.)

Genus 3. WITH A BASE OF SCHIST.

Species 1. PHYLLADE. (A name given in concert with Messrs. *Brochant* and *D'Aubuisson.*) Mixed thonschiefer of the German mineralogists. Different schists.

Base of clay-slate containing disseminated mica, quartz, felspar, hornblende, chiastolite, &c. either separately or together. The structure slaty.

Principal varieties.

GLANDULOUS P. ...... The crystals more or less well formed, rather equally disseminated, and enveloped in a phyllade, which is commonly micaceous.

Porphyritic. With crystals of felspar, &c.

(Environs of Anger. Deville and Laifour, department of the Ardennes, *Omalius d' Halloy.*)

Quartzose. With grains of quartz.

(Banks of the Mayenne near Angers.)

Chiastoliteferous. With crystals of chiastolite.

(Alençon, Tourmalet, Comelie, &c. in the Pyrenees. Burkhartswald in Saxony.)

MICACEOUS P......... Mica more or less abundantly disseminated in a phyllade, without any other accessory mineral, neither staining, nor whitening by the fire.

Platy. The mica disseminated in distinct and abundant plates.

(The greater part of greywacke slates. Planitz in Saxony. Combe de Gilliarde in Oisans.)

Satiny. Mica in scarcely distinct plates, the lustre silky.

(Schneeberg. Tourmalet in the Pyrenees.)

Dull. The mica disseminated in rare plates, aspect dull.

(Viel-Salm, department of the Ourte &c.)

CARBURATED P....... Slightly micaceous, stains black, and is sometimes calcariferous.

(Bagnere de Luchon. Hermersdorf in Saxony. Hofnungstolle, in the Hartz. Some marly bituminous schists of Thuringia.)

Species 2. CALSCHISTE.

Argillaceous slate, often predominant, and limestone disseminated either in elongated patches, small veins, or thin plates, sometimes parallel, sometimes traversing. The structure slaty.

Obs. This mixture of limestone and argillaceous slate is too constant in its structure, proportions, and character to be regarded as accidental, or as an argillaceous slate traversed by calcareous veins. The schist and limestone alternate in this rock in the same manner as the quartz and mica alternate in mica schist, the felspar and mica in gneiss, &c. I shall cite as examples of this rock: the veined calschiste of la Madeleine, near Moutier, which is micaceous, and its structure fissile, alternating, and fibrous; and those which I have observed at Mont Aventin, valley of the Arboust: at Lauderville, valley of Louron: and at the Pic d'Eredlitz, in the Pyrenees. They are blackish grey, micaceous, satiny, veined, and with a reticulated amygdaloidal structure, &c.

Genus 4. WITH A BASE OF TALC.

Species 1. STEASCHIST. (*Talkschiefer, Werner.*)

A talcose base containing disseminated mica or other minerals. The structure slaty.

Principal varieties.

ROUGH S. (Verharterertalc.) Generally brilliant, rough to the touch, mixed with petrosilex in laminæ, mica, disseminated pyrites, &c.
(Pesey, in the department of Mont Blanc.)

PORPHYRITIC S....... Disseminated nodules or crystals of lamellar felspar
(Vereix, in the valley of Aosta.)

NODULAR S........... Enveloped shapeless nodules of crystalline quartz, felspar, &c.
(The harbour at Cherbourg. Mont Jovet, department of the Doire.)

STEATITIC S.......... Soft, very unctuous to the touch.
(The stone of Baram, - St. Bel, near Lyon,—Dax.)

CHLORITE S.. .... Soft, green, mixed with chlorite.
(Corsica, with octohedral crystals of oxidulated iron. Cauteret.)

DIALLAGE S.......... Green or brown, mixed with diallage.

OPHIOLINE S. .... Mixed with serpentine. (Corsica.)

PHYLLADIEN S........ Talc and phyllade, very fissile.
(The gangue of the Valorsine conglomerates.)

Genus 5. WITH A SERPENTINE BASE.

Species 1. OPHIOLITE.*

A paste of serpentine enveloping disseminated oxidulated iron and other accessory minerals. The structure compact.

Principal varieties.

FERRIFEROUS O....... With disseminated grains of oxidulated iron.
CHROMIFEROUS O. .... With disseminated grains of chromate of iron.
DIALLAGIC O. .... With disseminated diallage.
(Baste, in the Hartz.)
GARNET O. .... With disseminated garnets.

Genus 6. WITH A LIMESTONE BASE.

Species 1. CIPOLIN.†

A base of saccharine limestone containing mica as a constituent and essential part. The structure saccharine, often fissile.

(Pyrenees.—Schmalzgrube in Saxony.)

Species 2. OPHICALCE.

Base of limestone with serpentine, talc, or chlorite. The structure imbedded, (Structure empâtée.)

Principal varieties.

RETICULATED O........ Egg-shaped nodules of compact limestone pressed against each other, and united as it were by a net-work of talcose serpentine.
(Marbre de Campan. Furstenberg in the Hartz.)

VEINED O............Irregular patches of limestone, separated and traversed by veins of talc, serpentine, and limestone.
(Vert Antique. Vert de Mer, Vert de Suza:)

GRANULAR O......... Talc or serpentine disseminated in a saccharine limestone.

---

* The greater part of common serpentines, potstones, &c. Noble serpentine should constitute a single mineralogical species.

† In order to shew the difference we make between pure saccharine limestone, and the rock with a limestone base which we name Cipolin, we shall state that cipolin often occurs subordinate to saccharine limestone.

Species 3. CALCIPHYRE.*

Paste of limestone enveloping crystals of different kinds. The structure imbedded, (empatée.)

Principal varieties.

FELSPATHIC C.........Crystals of felspar disseminated in a compact limestone.
(The Col du Bonhomme, *Brochant.*)

GARNET C............Garnets disseminated in a compact or saccharine limestone.
(The environs of the Pic du Midi in the Pyrenees.)

HORNBLENDE C....... Hornblende disseminated in a compact limestone.
(Isle of Tirey,† Hebrides.)

Genus 7. WITH A BASE OF CORNEAN. (Cornéene.)

Species 1. VARIOLITE. (*Blatterstein*, *perlstein*, some *mandelstein*, &c.

Paste of cornean containing nodules and veins, either calcareous or siliceous, which are contemporaneous or posterior to the paste.

Principal varieties.

COMMON V........... A compact paste of a sombre green, reddish brown or violet colour, with crystallized calcareous nodules.
(Variolites of the Drac, Oberstein, of the Hartz, &c.)

BUFONITE V.......... A black paste with calcareous nodules.
(Toadstone of Bakewell in England.)

---

* The division of the mixed rocks with a calcareous base into three species, far from being superfluous, is perhaps not carried far enough. Here even the characters of geological position accord with the mineralogical characters in indicating this division.

† Dr. Mac Culloch (Description of the Western Islands of Scotland, vol. 1. p. 50,) mentions that the Tirey limestone " is most distinguished for the quantity of augite dispersed through it," and states that the hornblende occurs occasionally in large concretious. (Trans.)

ZOOTIC V. .... Portions of entrochi mixed with the calcareous nodules.
(Kehrzu, near Clausthal in the Hartz, *de Bonnard.*)

VEINED V. .... Veins and small grains of spathose carbonate of lime.
(*Schaalstein* of Dillenbourg.)

Species 2. VAKITE.

Base of wacke containing mica, augite, &c.

Genus 8. WITH A BASE OF HORNBLENDE.

Species 1. AMPHIBOLITE.

Base of hornblende containing different disseminated minerals. The structure sometimes compact, sometimes fissile.

Principal varieties.

GRANITIC A. .... The structure compact, the texture granular, containing garnets, serpentine, bronzite, &c.

ACTYNOLITIC A. .... The structure compact, the texture saccharine, the colour green, containing garnets, &c.
(Kaf in Bareuth.)

MICACEOUS A. .... Hornblende and mica. The structure granular.

SCHISTOSE A. .... The structure fissile, the texture slightly fibrous.

Species 2. BASANITE.

A base of slightly brilliant and compact basalt, containing different disseminated minerals.

Principal varieties.

COMPACT B. .... Hard, compact, containing augite, olivine, titaniferous iron, &c. (Basalt properly so called.)

CELLULAR B. .... Hard, cellular. Egg-shaped cells rare.

Species 3. TRAPPITE.

A base of hard, compact, dull, and often splintery (fragmentaire) cornean trap, enveloping mica, felspar, &c. (Trap rocks).

Species 4. MELAPHYRE. (Trapporphyr, *Werner.*)

A paste of black petrosiliceous hornblende containing felspar crystals. Fusible into a black or grey enamel.

Principal varieties.

Demi-deuil M. ...... Deep black with white crystals, & no quartz. (Venaison in the Vosges. Sweden.)

Sanguine M. .... Blackish, the crystals reddish, with grains of quartz. (Niolo in Corsica.)

Green spotted M.... Reddish brown, with green crystals. (Black antique porphyry.)

Genus 9. WITH A BASE OF HORNBLENDE PETROSILEX.

The base is petrosilex coloured by hornblende, which is as it were dissolved in it: it is not black.

Species 1. PORPHYRY. (*True porphyry. Hornstein porphyr. Werner.*)

Paste of red or reddish* petrosilex, containing determinable crystals of felspar. Fusible into a black or grey enamel.

Principal varieties.

Antique P. .... Paste of a very deep red colour, with small crystals of white compact felspar.

Reddish brown P..... Paste of a reddish brown colour, with a little quartz. (Planitz,—Kusseldorf. Lesterel.)

Roseate P. .... Paste of a pale red colour, numerous grains or crystals of quartz. (Kunnersdorf in Saxony.)

Violet P.

Syenitic P.

Species 2. OPHITE. (*Green porphyry.*)

Paste of green hornblende petrosilex, containing determinable crystals of felspar.

* Leucostine, *Delametherie.*

Principal varieties.

ANTIQUE O. .... Green compact, homogeneous, opaque paste, with crystals of green felspar.

VARIED O. .... Brownish green granular paste with crystals of white, grey, or green felspar.
(The Tourmalet in the Pyrenees. Bode in the Hartz. Niolo in Corsica.)

Species 3. AMYGDALOID. (*Mandelstein, Werner.*) Some rocks improperly named variolites.*

Paste of petrosilex, containing round nodules of petrosilex, of a different colour from that of the paste.

Principal varieties.

GREENISH A. .... The tint generally green.
(La Durance.)

GREY A.

RED A.

PORPHYRITIC A. ...... A reddish paste containing small crystals of felspar and hornblende, with nodules.
(Orbicular porphyry of Corsica.)

Species 4. EUPHOTIDE. (*Haüy. Diallage rock.*)

Base of jade, petrosilex, or even of felspar, with numerous crystals of diallage. The structure granular.

(Corsica. Genoa, &c.)

Genus 10. WITH A BASE OF PETROSILEX, OR OF GRANULAR FELSPAR.

Species 1. EURITE. (*d'Aubuisson.* Some *weissteins. Klingstein. Werner.*)

Base of rather pure petrosilex, containing mica and other disseminated minerals. Structure either granular, fissile, or imbedded.

---

* Not only do the amygdaloids, as we here characterize them, differ from variolites in the nature of their paste, in that of their nodules, and in the relations of the formations of the two positions; but we moreover see that they differ in their geological position. Our amygdaloids are generally of a much more ancient formation than the variolites. All therefore tends to separate these two species of rocks which have been so much confounded.

Principal varieties.

COMPACT E........... The structure compact with mica and garnets,—no distinct crystals of felspar.
(Klingstein, *Werner*,—The rock of Sanadoire in Auvergne,—Coasme near Rennes.)

SCHISTOSE E.......... A fissile structure with a dense texture, &c.

PORPHYRITIC E....... Determinable crystals of either felspar or hornblende disseminated in the paste.
(Some hornstone porphyries,—flötz trapp-porphyr.)

Species 2. LEPTENITE. (*Haüy*. Some *weissteins*,—*Hornfels*, *Werner*.)

Base of granular felspar, containing mica and quartz as essentially constituent parts. The structure granular.

Species 3. TRACHYTE. (*Haüy*. A kind of porphyry.)

A fusible petrosiliceous paste of a dull aspect, enveloping crystals of vitreous felspar.

(Porphyritic rocks of the Drachenfels in the Sieben Gebirge. Mont d'Or.)

Genus 11. WITH A CLAYSTONE (*argilolite*) BASE.

Species 1. ARGILOPHYRE. (*Thonporphyr*, *Werner*.)

A claystone base enveloping crystals of compact or dull felspar.

Species 2. DOMITE. (*Von Buch*. *Lavas* of some mineralogists.)

Paste of harsh claystone, enveloping crystals of mica, &c.

(The Puy de Dome,—the Puy-Chopine in Auvergne,—the Isles Ponces.)

Genus 12. WITH A BASE OF PITCHSTONE, or OBSIDIAN.

Species 1. STIGMITE. (*Pitchstone* or *obsidian porphyry*.)

A paste of obsidian or pitchstone containing grains or crystals of felspar.

Genus 13. WITH AN UNDETERMINED BASE.

Species 1. LAVA.

A mixed or undetermined base, having evidently been melted, often porous, with cavities for the greater part empty, enveloping different minerals.

Principal varieties.

BASALTIC L. .... Black compact paste, with void cells which are more or less abundant.

TEPHRENIC L. .... The paste of an ash-grey colour, harsh to the touch, and porous.
(Tephrine, *Delametherie.* Lava of Volvic.)

SCORIACEOUS L. .... A black, grey, or reddish paste, with numerous vesicles, &c

PORPHYRITIC L. .... A vitreous or slightly lamellar paste enveloping crystals of vitreous and fibrous felspar.

PUMICE L. .... A pumice paste enveloping vitreous felspar.'

CLASS III. THE AGGREGATED ROCKS.

Character. Formed by mechanical aggregation; the cement or paste posterior to the parts it contains.

Genus 1. THE CEMENTED ROCKS.

The parts united by a slightly apparent cement.

Species 1. PSAMMITE. (Micaceous sandstone,—coal grit,—the greater part of greywackes.)

A granular rock, principally composed of small grains of quartz mixed with other minerals, and united by a slightly apparent cement of a different nature.

Principal varieties.

QUARTZOSE Ps. .... Middling sized grains of quartz are essentially predominant, with some disseminated grains of felspar, &c.
(Remilly, near Dijon. Martes de Vayre, near Clermont in Auvergne,—above Carlsbad in Bohemia.)

GRANITIC Ps. .... Distinct grains of quartz and felspar in nearly equal quantities, connected almost without cement.

(Chateix, near Royat. Mont Peyroux in Auvergne.)

MICACEOUS Ps........ Grey sandy paste, containing numerous plates of mica.

(The greater part of the coal grits.)

REDDISH Ps........... A reddish sandy paste mixed with mica.

(Micaceous red sandstone.)

(The heights of the environs of Saarbruck, &c.—Athis, near Feugeurolle, in the environs of Caen. Rothe-todt-liegende of Vaterstein, near Henstadt, in the Hartz.—Kaufinger-Wald, near Cassel.)

SCHISTOSE Ps. .... A blackish argillo-sandy paste, containing mica.

(The greater part of schistose greywackes.)

CALCAREOUS Ps....... A tolerably conpact sandy calcareous paste with mica.

(Bonneville, near Geneva.— Lautenberg in the Hartz.—Hauszelle, near Zellerfeld, in the Hartz.)

Genus 2. THE IMBEDDED ROCKS.

Fragments enveloped in a very distinct paste.

Species 1. MIMOPHYRE. (Some greywackes.)

An argillaceous cement uniting very distinct graius of felspar, and sometimes of quartz, clayslate, &c.

Principal varieties.

QUARTZOSE M........ Hard and solid, with numerous grains of quartz.

(Châteix, near Royat in Auvergne,—Summit of the Pormenaz, in the Savoy Alps,—near the Vallorsine conglomerates.)

ARGILLACEOUS M. .... Friable with some grains of quartz and mica, with fragments of carburated schist, &c.

(Flöhe, between Freyberg and Chemnitz; the argillaceous paste is green with small rose-coloured crystals of felspar. The red *thonstein* with white spots of Zaukerode, near Tharand.)

Species 2. PSEFITE. (The greater part of the *todt-liegende.*)

An argillaceous paste enveloping disseminated and midling sized fragments of mica-slate, clay-slate, whetstone-slate, and other rocks of the same formations.

Principal varieties.

RED Ps. .... With a red paste.
(Fragments of whetstone slate, grains of felspar, &c. rothe-todt-liegende of Zorge in the Hartz,—with small grains of quartz, rothe-todt-liegende of Elrich in the Hartz, —fragments of mica slate, clay slate, &c. thonporphyr of Clemnitz in Saxony.)

Species 3. PUDDINGSTONE. (*Poudingue.*)

A rock principally composed of tolerably large uncrystallized portions, agglutinated by a paste.

Principal varieties.

ANAGENIC P. .... Primitive rocks united by a cement, either schistose or of saccharine limestone.
(Trient in the Valais. Col de Cormet, in the late department of Mont Blanc.)

PETROSILICEOUS P.....Rocks of all kinds united by a petrosiliceous cement.

ARGILLACEOUS P...... Quartzose nodules united by an argillaceous cement.
(Lautenthal in the Hartz.)

POLYGENIC P......... Rocks of all kinds united by calcareous cement.
(Nagelfluhe of the Rigi.)

CALCAREOUS P........ Calcareous nodules united by a calcareous cement.
(Nagelfluhe of Salzbourg.)

SILICEOUS P,......... Nodules of silex in a paste of homogeneous sandstone.
(Environs of Nemours.)

JASPER P............ Nodules of agate, &c. in a paste of agate or jasper.
(Pebbles of Rennes.)

PSAMMITE P.......... Nodules of silex, &c. in a psammite paste.
(Scotland, employed at London in the construction of the docks.)

Species 4. BRECCIA.*

A rock principally composed of angular, uncrystallized middle sized fragments, agglutinated by a paste.

Principal varieties.

QUARTZOSE B.........Fragments of quartz and other rocks united by a serpentine paste.
(Col de Queyriere, in the Brianconnais.)

SCHISTOSE B.......... Fragments of schist, phyllade, &c. in an argillaceous paste.
(Todt-liegende of Eisenach.—The coast near Saint Jean de Luz.—Coutance.)

SCHISTO-CALCAREOUS B. Fragments of schist or other argillaceous rocks, in a more or less calcareous paste.
(Environs of Elbingerode in the Hartz.—Braunsdorf in Saxony.)

CALCAREOUS B. ...... Calcareous fragments in a calcareous paste.

VOLCANIC B. ....Fragments of pyrogenic (pyrogènes) rocks enveloped in a calcareous, argillaceous paste of wacke, lava, &c.
(Aurillac. Gergovia. Habichtswald in Hesse. —Rome.)

* There are no precise limits between some puddingstones and some breccias, but there are so many important differences between siliceous puddingstone and calcareous breccia that we can, on no account, unite these two rocks in the same species.

CORRIGENDA.

page	*line*	
4	5	from the bottom, *for* 5 inches *read* 11 inches.
17	4	from the bottom, *for* in *read* amidst.
57	3	*for* is *read* are.
67	5	from the bottom, *for* regret *read* reject.
80	3	*for* that are *read* which is.
95	14	*for* böhlen *read* höhlen.
108	17	*dele* only.
137	2	*for* veins *read* vines.
ib.	7	from the bottom, *for* real *read* coal.
141	16	*for* rock *read* rocks.
158	6	*for* villages *read* valleys.
162	3	*for* state of relative *read* study of respective.
ib.	21	*for* are either *read* is either.
167	10	from the bottom, *for* foundations *read* situations.
171	10	*for* on *read* in.
181	5	*for* so *read* to.
186	5	*for* graphite *read* gryphite.
ib.	16	*for* exclusively *read* inclusive.
187	3	from the bottom, *for* in *read* is.
196	9	*for* possess *read* partake.
198	12	*for* rocks *read* works.
203	13	*for* formation *read* formed.
223	9	*for* into Nahe *read* into the Nahe.
211	3	from the bottom, *for* 3 feet 1 in. *read* 13 feet 1 inch.
213	6	*for* feet *read* inches.
214	9	*for* feet *read* inches.

PLATES.

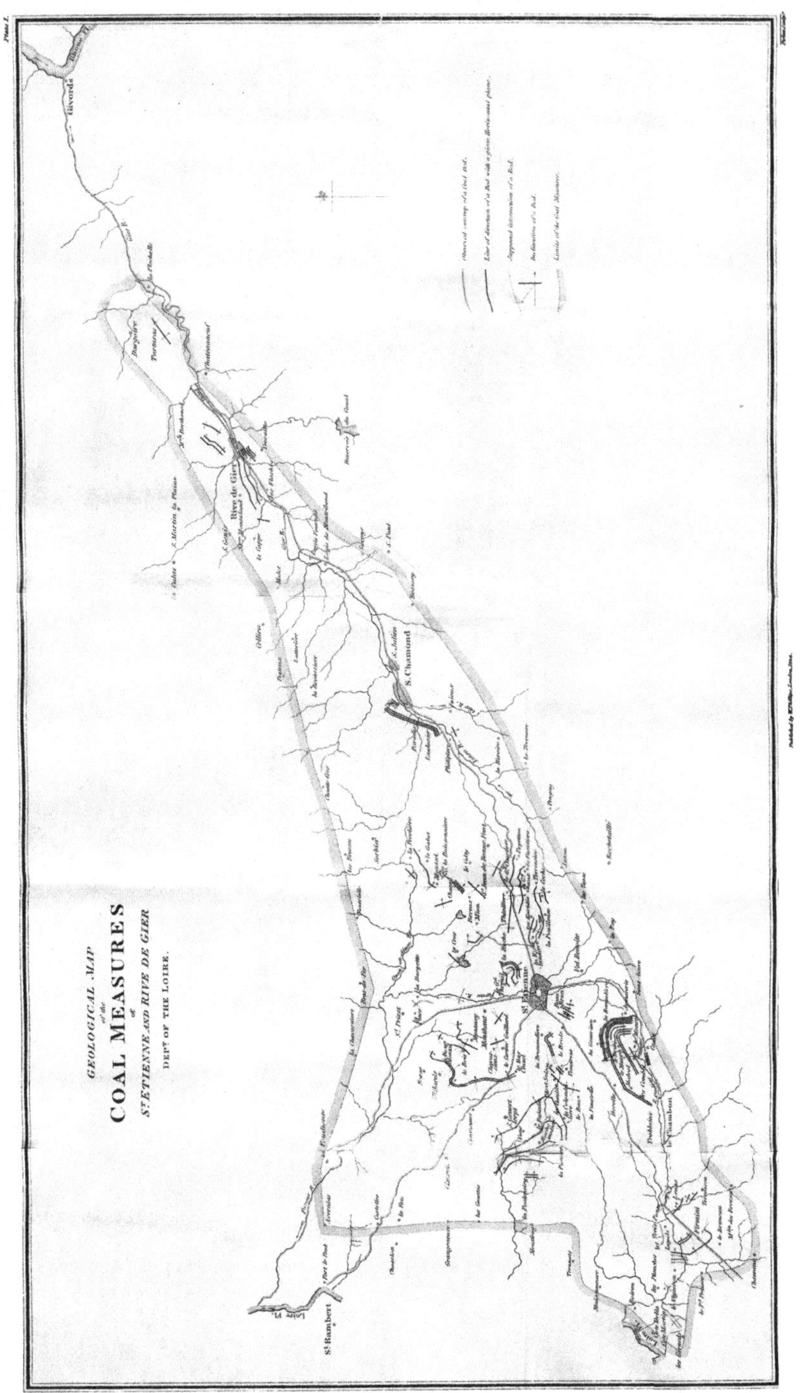

The material originally positioned here is too large for reproduction in this reissue. A PDF can be downloaded from the web address given on page iv of this book, by clicking on 'Resources Available'.

Geological Sketch of the Coal District of Saint-Etienne, department of the Loire. By M. Beaunier, Engineer in Chief of the Royal Mining Corps.*

[Annales des Mines, 1816.]

THE coal district of the arrondissement of St. Etienne is contained in every direction within a basin of primitive rocks, extending from SW. to NE. between the Loire and the Rhone, towards the points, where those two rivers, flowing in contrary directions, are nearest to each other. See Plate I.

This basin swells out considerably; towards the Loire on the west, its greatest breadth, taken in the meridian of Roche-la-Moliere, is 13,000 metres [42,652 feet]; its edges approach each other towards Saint-Chamond, and follow thence both sides of the river Gier, parallel with its course to the Rhone, eastwards to the limits of the department of the Loire: the basin is continued without sensibly altering its direction to and even a little beyond † the Rhone.

At Rive de Gier the coal formation is not more than 2,300 metres [7,546 feet] in breadth, and at Tartaras it is still less. Its greatest length, measured from St. Paul de Cornillon (on the Loire) to Givors (on the Rhone), is 46,250 metres [151,742 feet]. It covers a surface of 221.43 square kilometres [about 87 square miles]. When speaking of the limits of the coal basin, a continuous line is understood, which would mark the separation on the surface of the coal

* This Sketch forms part of a detailed account of the manner of working the coal mines &c. in the above district. (Trans.)

† Traces of coal are found at Ternay, on the left bank of the Rhone.

measures from the primitive rocks on which they rest, if they were both always exposed. The suite of elevations, consisting of primitive rocks, rising above the coal district, might be described by the same terms.

Towards the south these heights belong to the principal chain of the Pilat, separating the department of the Loire from that of the Ardèche, the ramifications of which towards the SW. form the limit of the department of the Haute-Loire. On the west, the elevated primitive country rises above the right bank of the Loire, for a short distance, commencing from St. Paul de Cornillon, and afterwards forms in the direction of the Rhone, a nearly continuous ridge on the north of the coal basin, in the same manner as the Pilat chain rises above it on the south and south west.

The principal part of these primitive rocks is composed of granite, the most abundant ingredients of which are felspar and mica. The first of these substances most frequently presents itself in the form of a nucleus entangling plates of mica, and occasionally talc. It is not rare however to find the different ingredients of the granite more distinct; the mica is less abundant, and the crystals of felspar (frequently of a roseate tint) become of considerable size. In numerous places mica is the most abundant ingredient, the rock then becomes a gneiss, and may be considered sometimes as a *granit veiné*, and sometimes as a mica or talcose slate.

On the west and north west the coal measures rest in general immediately on granite; on the south and south west, ordinarily on gneiss, mica, or talcose slates, or even upon serpentine; these rocks separate them from granite, which is found on approaching the primitive ridges. This primitive country contains metalliferous veins.*

That we may not pass the limits we have prescribed ourselves, we will now proceed to consider the country formed of the coal-measures.

* Mines of sulphuret of lead at St. Julien, Molin, Molette, and the S.E. of Pilat; several veins of the same substance on the NW. of Rive de Gier, towards Fontain, and St. Martin-la-Plaine.

The latter are composed, if the expression may be allowed, of the more or less divided debris of the basin that contains them; these debris are disposed in beds of variable appearance alternating with others of coal and shale, which, besides the greatly triturated debris of which they are composed, contain vegetable remains in different states of preservation.

The following is a list of conglomerates (Poudings), sandstones, shales, and coal beds, filling up the coal basin.*

1. Conglomerates formed of large fragments of primitive mica and talcose slates and granite, scarcely cemented together; these fragments are often of several cubic metres content. (Fine specimens on the north of St. Etienne towards Fouillouse; at Rive de Gier, between that town and the bridge of la Madeleine, &c.)

2. Conglomerates composed of smaller sized fragments, cemented together by the ordinary sandstone of the coal measures. (These conglomerates are of more frequent occurrence than the last).

3. Large grained sandstone mixed with small rolled fragments of different rocks.

4. Compact sandstone, the grains of an uniform size, containing plates of mica. (This is the building-stone known in the country by the name of Molasse). It forms beds of considerable thickness.

5. A fine grained sandstone, in thin beds.

6. A fine grained slaty micaceous sandstone, the slates being very thin.

7. A friable micaceous slate, in which small grains of sand are distinguishable.

8. A more compact schist, in which mica is still visible.

9. A harder schist. This kind is the least common.

10. Coal, containing little bitumen, earthy, and mixed with schist. (This bed is found at Tartaras, Saint-Chamond, and at Rive de Gier, and is known by the name of bâtarde.)

* This enumeration agrees in many respects with that given by M. de Bournon, in his Essai sur la lithologie du Forez.

11. Coal, more bituminous than the last, homogeneous, with a brilliant fracture, generally hard, and separating in large fragments. (This is the coal most esteemed for fuel; it is known at Rive de Gier by the name of *Raffand*, and occurs abundantly in the vicinity of St. Etienne, that in the greatest repute comes from la Berandiere or la Ricamerie).

12. Very bituminous coal, very homogeneous, of a fine black colour, with a brilliant fracture, and slightly friable, this is the variety known in the country by the name of *Maréchale*. (The most esteemed coal of this kind occurs in the districts of Grand Croix, and la Chanchère, near Rive de Gier, and St. Etienne; in that of the Bois d'Aveize, of Clusel, and of Roche-la-Molière, upon the bed named Seignat; this variety is generally reserved for forges).

The conglomerates, composed of large fragments scarcely cemented together, never immediately accompany the coal; wherever they exist, they form the lower beds of the coal measures, resting on primitive rocks; frequently however the lower beds, in immediate contact with the primitive country, are of different kinds of sandstone, upon which, near the edges of the basin, beds of coal rest.* When these beds occur of little extent, they are generally contained between others of the compact sandstone, variety No. 4.

The coal often rests immediately on this sandstone; at other times (and this case is rarer here than elsewhere) the roof or floor of the coal bed, or both together, are of schist, containing vegetable impressions. There are few coal beds that are not divided in their thickness by seams of a more or less compact schist, bearing in the country the name of Gore.

It is impossible to say any thing on the general thickness of the coal beds. In certain situations † coal not more than 48 centimetres [about 5 inches] thick is worked. Ordinarily the workings are carried on in those varying in mean depth

* A fine example of this fact is seen on the road from Rive de Gier to Lyon, above the bridge of la Madeleine.

† At Rive de Gier, for example, at the colliery of Mont Dixien.

from 1 to 5 metres [about 3 feet 4 in. to 16 feet 8 in.] in different points these same beds swell out suddenly, so as to acquire a thickness of from 16 to 20 metres [about 53 feet 4 in. to 66 feet 8 in.]; or else not less suddenly they diminish in thickness from the floor and the roof approaching each other, to a point at which no traces of them are found for a considerable distance. The latter accident, known in the country by the name of coufflée, occurs more frequently than the other in the coal measures under consideration—this circumstance occasions great inconvenience in the working of St. Etienne collieries, as it throws a great difficulty in the way of tracing the continuity of the coal beds.

Having thus shewn the most ordinary superposition of the beds in the coal measures, it is necessary to consider the common relations of the different masses. For this purpose that, which occurs when the basin forms a single valley, must be distinguished from that which happens when the basin, greatly dilated, is cut through in different directions by several small valleys.

Where the coal formation is confined between two continuous and parallel chains, the beds are trough-shaped, and form a new valley enclosed in that of the primitive country. This is the case in the greater part of the Rive de Gier district, the beds are conformable to the sides of the valley in which they rest, being horizontal or slightly curved at the bottom, and rising rapidly on either side.

When, on the contrary, and as is generally the case on the watershed towards the Loire, the coal formation is considerably dilated, and that it is cut in different directions by valleys of greater or less depth, it is observable with but few exceptions, that all the beds of the coal measures are inclined in a different direction from the slope of the little isolated hills or terraces formed of them. Thus the outcrops of the beds generally form belts round the hillocks or platforms, and appear on the map as winding lines, differing little in level.

The coal beds then can, relatively to the portion of the coal measures that occupies us at present, be considered as curved surfaces, tending to form small basins, whose inferior parts, that is to say, the points furthest removed from the surface, are placed precisely under those parts of the coal formation that are at present the highest; a fact that has been observed in many other countries, but from which the singular and rigorous conclusion has not perhaps been drawn, that the lowest part of the primitive rocks on which the coal formation is deposited, answers precisely to those parts of the latter, which are at present the most elevated; or reversing the proposition, that the last valleys cut in the coal measures cover ridges of primitive rocks concealed beneath them. This sketch, which is perhaps new, supposes certainly what many geologists are inclined to doubt, namely, that the beds in the coal formation are at present in the same situation as when they were moulded in the primitive rocks.*

This supposition seems to us supported, with respect to the country in the vicinity of St. Etienne, by so many facts, that we have not been able to refuse it credit—we shall present the following considerations.

1. When the coal beds, or the beds of the measures, are highly inlined, it generally happens that their thickness increases with their depth. An analogous effect would take place with regard to matters deposited on inclined planes.

2. When the lowest deposits of a secondary country have been formed by large fragments with little or no adhesion, it follows that these matters have rolled into the hollows of the primitive country, so as partly to fill them up, forming a less inclined talus upon which beds of sandstone, schist, and coal have been deposited, which then would naturally be

* Although it does appear that the irregularities of St. Etienne coal measures may be owing to the uneven surface of the primitive rocks beneath, yet it is almost needless to remark that faults and many contortions must have originated from other causes; for example, no possible inferior surface on which the coal measures could have been deposited, could at all have formed the twistings and contortions of the coal measures from Broad Haven to Gouldtrop Road, in St. Bride's Bay; they apparently arise from the trap, which rising from beneath the old red sandstone, overflows the coal measures and squeezes them up against the greywacke. I have forwarded sections, &c. of this remarkable coast to the Geological Society in an account lately drawn up by Mr. Conybeare and myself on the Geology of Southern Pembrokeshire. (Trans.)

much less inclined than the declivities of the primitive rocks near them. Thus, multiplied observations prove, that this state of things which can easily be imagined, has not been distended by any catastrophe subsequent to its formation.

3. In the numerous places where the immediate superposition of the coal measures on the primitive country can be observed, the former is clearly seen to take all the curvatures required by the uneven surface of the latter, without any loss of continuity or fracture of the beds. Would appearances be thus, if the rocks anterior to the deposit, or the deposit itself, had been subjected to shocks and disturbances?

4. A single coup d'œil thrown over the atlas accompanying the work from which this memoir is extracted, shews that, in the widest part of the basin, the coal measures have entangled the elevated points of the primitive country, it would be easy by following up all the details, to give nearly the configuration of the rocks at present concealed, and to conclude that no important change has altered it.

We could greatly extend these considerations and fortify our opinion by a greater number of facts, did not the fear of passing the bounds we have prescribed ourselves deter us. It is here the place to state, that the coal measures, surrounded on all sides by primitive rocks, support, but over a small extent, the remains of a newer formation, torn away at the time when valleys were last formed. Patches of it are found at present only on the elevations of St. Priest and La Tour, situated near the river Furens, between St. Etienne and the Loire.

This formation is remarkable, as volcanic tufa and fragments of basalt are embedded in a siliceous substance, which has been deposited in horizontal beds on the coal measures.

This fact does not appear elsewhere in any part of the St. Etienne district, the volcanic hillocks, found in the plain of Forez, have no immediate connexion with the coal measures under consideration.

M. de Bournon has described the minerals of St. Priest and la Tour, and given a detailed list of the different accidents that have affected them. He mentions well characterised siliceous fossil-wood.*

It is to be regretted that the youth of the author, and the state of science when he wrote, should have given rise to several mistakes, which it is not our present object to rectify.

The levels of the different rocks passed in review, can be ascertained from the atlas of the Loire mines.

1. The primitive country can be traced without interruption from the Rhone near Givors (at a height of 169 metres [563 feet] as far as the principal heights of the Pilat, elevated 1,200 metres [3,937 feet] also above the level of the sea.†

The pit of Logis des Pères near Rive de Gier, gives the deepest section of these coal measures, its depth is 25 metres [82 feet] beneath the level of the sea; ‡ on the other hand their greatest elevation is (at Mont Salson, near St. Etienne) 725 metres [2,379 feet] above the same level. It follows that the greatest observed depth of the coal measures is 750 metres, [2,461 feet.]

3. The greatest observed elevation of the formation at St. Priest and La Tour is 600 or 650 metres, [1,968 to 2,134 feet].

* Essai sur la Lithologie des environs de St. Etienne, p. 45.

† The height of the Pilat has been estimated at 1,215 metres [3,986 feet] in the Annuaire statistique du department de la Loire, printed by order of M. de Colombier. We do not know what observations led to this result.

‡ The depth of the coal measures 8 metres [261 feet] below the level of the sea, is shewn by the pit of Matouret, 325 metres [984 feet] in depth, the mouth of which is higher than that of Logis des Peres.

Sketch of a Geological Map *of the Paris*

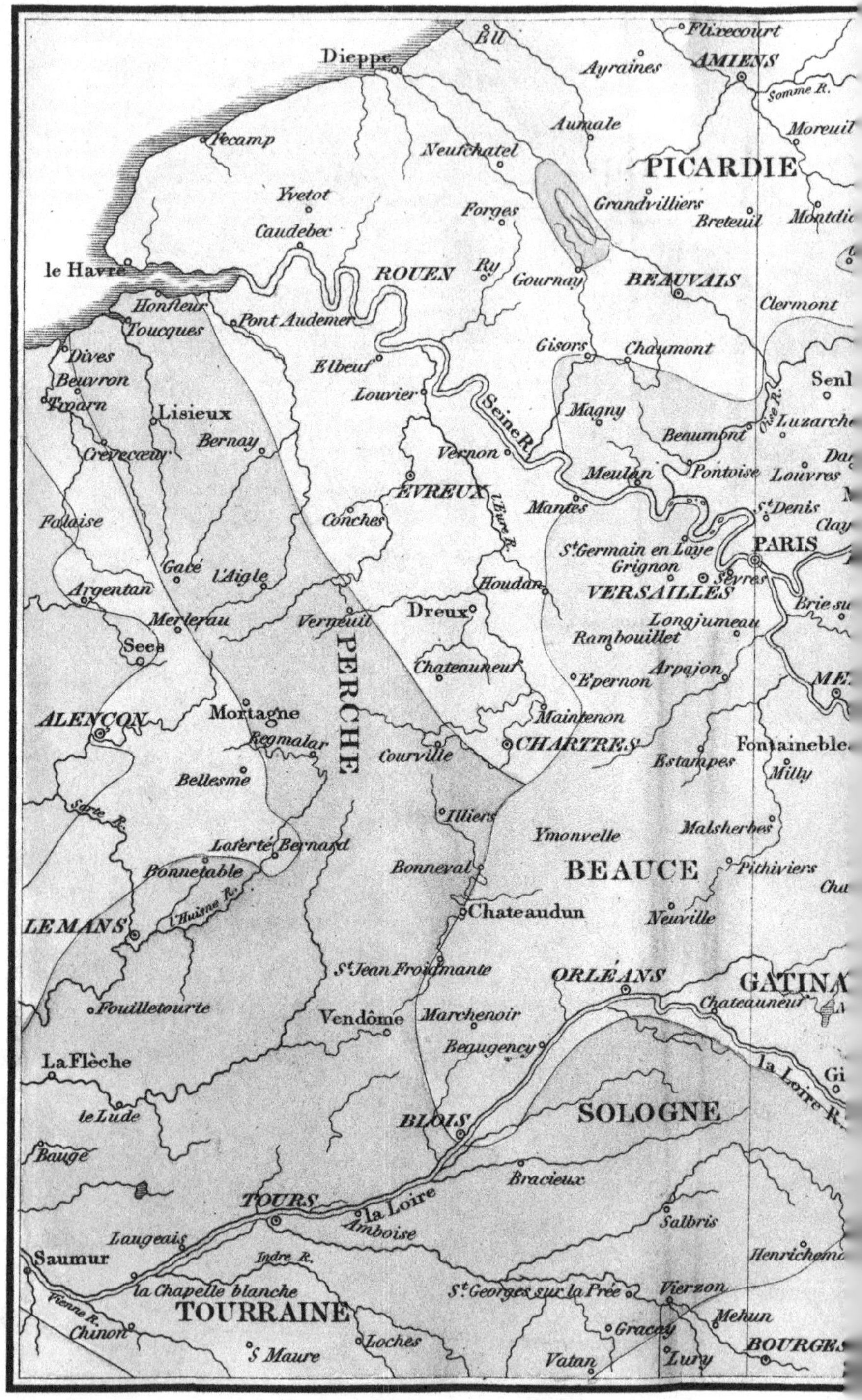

Basin, and the neighbouring Countries.

Published by W. Phillips. London. 1824.

Basin, and the neighbouring Countries.

Published by W. Phillips, London. 1824.

By J. J. d' Omalius d' Halloy.

Plate II.

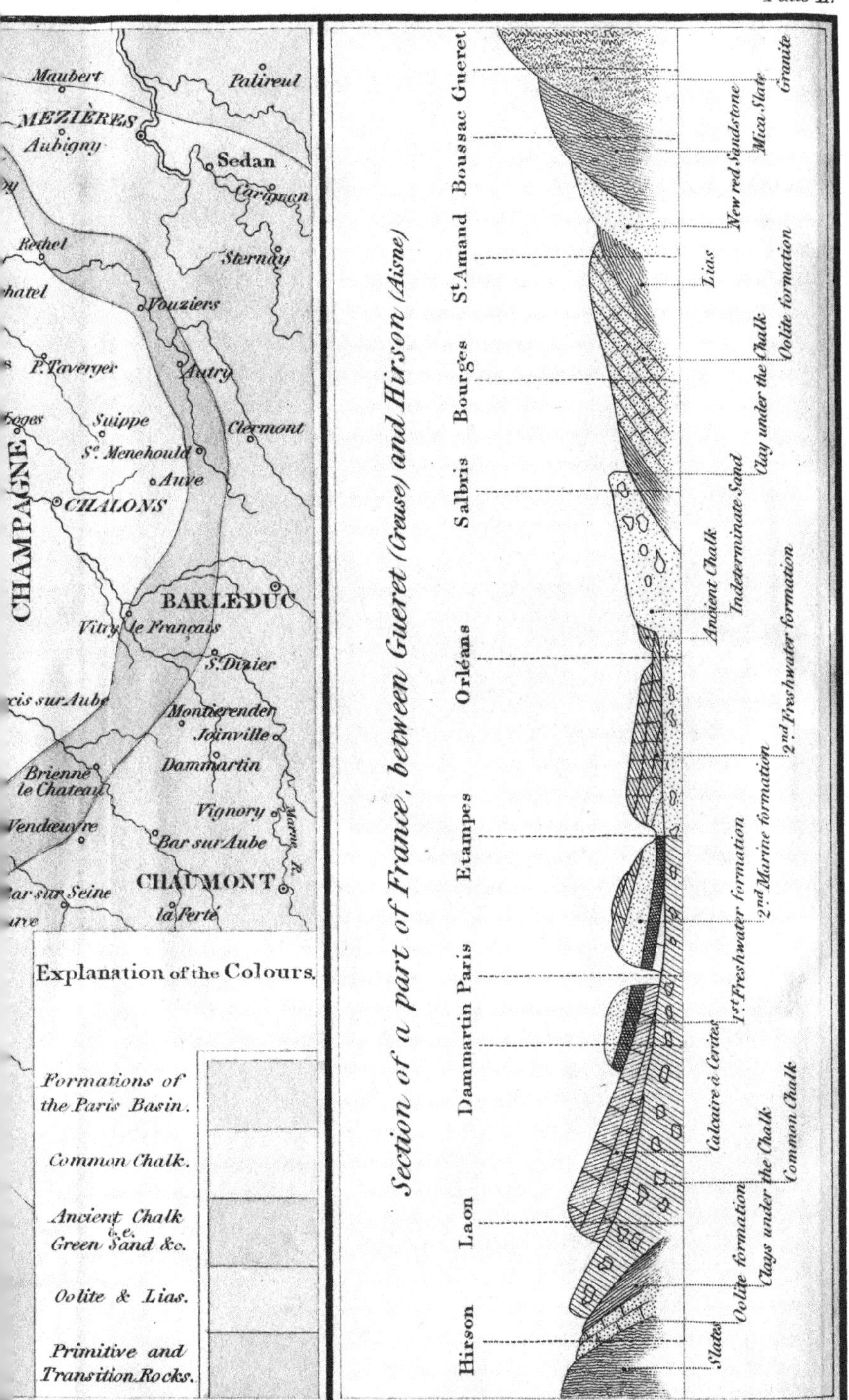

Pickett sculpᵗ.

Memoir on the Geographical Extent of the Formation of the Environs of Paris.—By J. J. D'Omalius D'Halloy.

Read at the Institute, August 16, 1813.

(Annales des Mines, 1816.)

THE learned researches of Messrs. Cuvier and Brongniart, have drawn general attention to the formations in the environs of Paris, and this is not remarkable; for if the lofty Alps, where nature presents herself under such magnificent forms, have inspired the great de Saussure, the true creator of geology as a science of observation; if Saxony, a country excavated to vast depths, in consequence of its mineral riches, has offered to the genius of Werner, an opportunity of establishing the first good system of geology; the environs of Paris, containing such abundant remains of living creatures, have given birth to true philosophical geology, that which drawing its conclusions from the knowledge of the organised bodies entombed in the bosom of the earth, can alone afford the certain means of comparison between distant formations, and will one day perhaps throw some light on the various catastrophes that have changed the surface of the globe, as it has already indicated the nature of the liquids in which some of these phenomena have occurred.

The geographical extent of the formations in the environs of Paris, and the details into which Messrs. Cuvier and Brongniart have entered in their geological map, have not allowed them to represent the entire limits of this formation.

I conceived that the determination of these limits throughout the extent of the basin would afford some interest, and I have with this design undertaken several excursions, the results of which I here present. I should however acknowledge that a part of this work has already been executed by M. Desmarest, sen., who had carefully determined the limits of the Chalk in Champagne. I have also received great assistance from the mineralogical atlas of M. Monnet, a work full of valuable observations, and less known than it deserves to be. Lastly, I have consulted two memoirs of Messrs. de Tristan and Bigot de Morogues, with advantage, for that part of this basin which occurs in the neighbourhood of the Loire.

The different formations of which the country in the vicinity of Paris is composed, taken collectively, a few isolated patches detached from the mass being omitted, occupy a surface of about 170 square myriameters, (7100 square miles English) forming an irregular polygon, elongated from north to south, whose greatest axis may be represented by a line, 30 myriameters long (328,391 English yards), drawn from Laon to Blois, the sides of this polygon pass in the vicinity of the towns of Laon, La Fère, Noyon, Clermont, Beaumont, Gisors, Mantes, Houdan, Chartres, Châteaudun, Vendôme, Blois, Orléans, Cosne, Montargis, Nemours, Nogent sur Seine, Sezanne, Epernay, and Reims. The Parisian formations rest throughout this extent upon chalk, which forms, as Messrs. Cuvier and Brongniart have remarked, a vast belt round the Paris Basin. (See Plate 2.)

These limits are very easily determined on the north of the Seine, being both physically and geologically distinguished; the Parisian formation every where presents itself under the form of a chain of hills more or less indented, which rise above the chalk plain. The latter becomes, however, lower and more even than it usually is, as it approaches the foot of these hills. Messrs. Cuvier and Brongniart have described numerous *chutes* of the Parisian rocks towards the chalk plain; but as they have not had occasion to speak of that near Damerie and Reims, I shall make a few remarks on it.

The chalk, hid on the east of Paris by posterior formations, begins to shew itself in the valley of the Marne below Dormans, and rises as the valley is ascended, so that arriving at the plain of Champagne this formation is seen constituting the base of the hills for some yards above the level of the plain. This fact, observable in many other places on the border of the Parisian rocks, proves that a part of the valley of the Marne has been hollowed out of chalk, and appears to shew that the existence of the low plain, bordering the Parisian hills, is not the effect of chance, giving this form to the surface of the chalk, before the deposition of the formations composing the hills, but that it owes its origin, in a great measure, to the same cause which has worn the exterior border of these hills in such a manner as to form those numerous capes, islands, and gulfs, which are there observable.

I have not seen the plastic clay formation in this part, but according to the observations of M. Desmarest, jun. it shews itself under the form of black earth, often sandy, sometimes clayey, and almost always impregnated with carbonaceous matter. This black earth, of which M. Desmarest proposes to publish an account, bears a great affinity to that worked for the preparation of sulphate of iron, and which is very common in the northern part of the Paris basin, and even in the chalky plain, constituting isolated deposits in the form of islands or small basins. The resemblance of this black pyritous earth to plastic clay, will no doubt throw new light upon the history of that formation, of which it spreads the extent, at the same time that the occurrence of fossils characteristic of the *cálcaire à cérites* in some beds of this black earth,* shews that the plastic clay bears a great affinity

* I have observed but two places where this earth contains cerithia. One is at St. Marguerite, near Dieppe, (Seine Inferieure), where it forms a small basin in the chalk formation. A series of clay and sandy beds is there seen, the first of which alternate with a few strata strongly impregnated with carbonaceous and pyritous matters, worked for the fabri-

to the calcaire à cerites, as Messrs. Cuvier and Brongniart have already suspected.

The calcaire à cerites formation,† which appeared to me to rest immediately on the chalk between Damerie and Reims, does not furnish any good building stones; its beds are on the contrary loose and friable, as at Grignon, and contain an immense quantity of shells. In this system of hills, are found the fossils, celebrated as those of Courtagnon, a name given them, because M. de Courtagnon was the first, in the Chateau of the same name, to form a considerable collection of these shells; for they are equally abundant, and easier to collect at Fleury, la Riviere, and Arthy, than at Courtagnon. These shells are known to be generally the same with those of Grignon; it could scarcely be expected that two deposits so distant, should so resemble each other

cation of sulphate of iron, and layers of shells more or less broken, among which are distinguished many cerithia, and bivalves which I believe belong to the genus cytherea.

The other situation is near Château-Thierry (Aisne), where the valley of the Marne affords blackish clay, full of shells, among which are seen many oysters, cythereæ, and a cerithium resembling that of St. Marguerite. It is very probable that this deposit is situated between the chalk and the calcaire à cerite, since this last forms the surrounding hillocks, and that the chalk is met with about 1000 yards further up the valley.

(Note of the Editor of the Annales des Mines.)

We visited this deposit in August, 1813; it evidently constitutes a part of the calcaire à cerite formation, and its inferior parts are composed, as the author has said, of pure plastic clay, and of plastic clay mixed with sand, lignite, pyrites, and the shells mentioned by the author; oysters, moreover, are there seen in great abundance. It should be observed that this disposition is precisely the same, and with the same shells, at Marly near Paris, above the chalk; at Vauxbuin, near Soissons, department of the Aisne, &c.

† Messrs. Cuvier and Brongniart do not mention the calcaire à cerites formation, in their enumeration of the Parisian rocks (2d edition, 1822) The following is the list given by them:

in this particular,* for, with few exceptions, the same species are distributed in a similar manner. But with respect to preservation, the shells at Courtagnon excel those of Grignon; they are harder, less white, and possess a pearly appearance, resembling recent shells. The calcaire grossier containing them has a yellow tint, approaching more the colour of ochre than at Grignon; in some places it is quite friable, in others

Formations.		*Sub-formations and principal Rocks composing them.*
SECONDARY ROCKS.		
I. Ancient marine formation	1.	Chalk.
TERTIARY ROCKS.		
II. First freshwater formation	2.	Plastic Clay. Lignite. First Sandstone.
III. First marine formation	3.	Calcaire grossier, and the Sandstone that it sometimes contains.
IV. Second freshwater formations	4.	Siliceous limestone.
	5.	Gypsum with bones. Freshwater marls.
V. Second marine formation	6.	Marine gypsum marls.
	7.	Third upper marine sandstone and sand. Upper marine limestone and marls.
VI. Third and last freshwater formation		Millstone without shells. Millstone with shells. Upper freshwater marls.

The calcaire à cerites appears to be included in the calcaire grossier; as the author always speaks of it under the former name, I have not thought myself at liberty to alter it.—(Translator.)

* It may be said that the deposit of shells observed at Grignon, owes its celebrity to the number, the beautiful preservation of the shells that it contains, and the facility of procuring them entire; but the bed of which this deposit forms a part is not wanting in any place where even patches of calcaire à cerites occur, and it perhaps extends to greater distances than is supposed. This is not the place to give proofs of this opinion; the Grignon bed is found not only in the Paris basin, but also on its outskirts, every time the calcaire à cerites is sunk through; this bed, it is true, will not forcibly strike observers, except when, being friable, the shells may be obtained entire, as at Perne; at Vivray, near Liancourt; at Mont Ouen, and Mont Javoult, near Gisors; at Septeuil, to the SW. of Mantes, &c.—(Note of the Editor of the Annales des Mines).

the particles are closely united; and as these different states of induration are irregularly disposed in the same mass, the hard parts might be said to be owing to the infiltration of a species of calcareous cement.* Above this sandy limestone, are beds of a compact, white, and solid limestone, alternating in the higher parts with greenish marl. I have not discovered any shells in place in this formation; I only found detached fragments of white limestone, containing the interior casts of cyclostoma mumia, and which I believe, from appearances, to belong to the inferior beds. A stone is employed for building in these cantons, particularly in the environs of Dormans, containing a great abundance of cerithium lapidum. It is a whitish fine grained limestone, a little cavernous, like the freshwater limestone, and which appears to me to come from the beds between the true calcaire à cerites, and that which contains the cyclostomæ; as it more resembles this last than the common marine limestone, I should be inclined to believe that it had also been formed in fresh water, and that these two beds belong to the same system as the clicart of Mantes la-Ville, described page 229 of the mineral geography of the environs of Paris.†

* These solid limestone nodules very frequently occur in the friable strata below the calcaire à cêrites, and are seen at Grignon; at the descent of Beaumont sur Oise; at Meudon; at Issy, &c. &c. They project beyond the vertical sides of these beds, and often appear as if penetrated by calcareous spar, which gives a radiant appearance to their fracture.—(Note of the Editor of the Annales des Mines.)

† I shall here observe that many geological circumstances, joined to zoological characters, make me believe that the cerithium lapidum ought to be arranged with the Potamides of M. Brongniart, or cerithia of the freshwater formations. This shell, differing very little from Potamides Lamarkii, appears to me to possess this peculiarity, that it occurs in the last marine beds, and in the first strata of the freshwater formation, and that it is the only fossil of the marine formation that *really occurs in place* in the freshwater formation.

I shall to this add that I have observed a potamides at Etampes which appears to me more slender, and with less defined tubercles than Potamides Lamarkii. I conceive that it may be considered as a separate species, or principal variety, that it may be named P. acuminatus. It occurs in the

The green marls are covered by another limestone equally white, but a little less compact than the preceding, presenting tubular cavities, characteristic of particular parts of the freshwater formation, and containing a great quantity of shells, among which may be particularly distinguished two species of lymneæ and a small amphibulimus, (Bulimus pusillus. Brong.)

Lastly, the whole is surmounted by the millstone without shells, with the sands and clays which usually accompany it, and which cover all the platforms in the vicinity.

This order of superposition, sufficiently worthy of attention from the series of beds that it exhibits, is still more remarkable for the constancy and uniformity with which it appears in the whole of the country extending from Chateau-Thierry to Reims.

I regard all the portion of this country placed above the calcaire à cerites, as belonging to the freshwater formations.

white limestone, forming the upper part of the freshwater formation; it is generally transformed into white silex, and often attached to nodules of that substance.

The true P. Lamarkii occur in the same situation, forming a species of deposit in the midst of a bed, whose upper part consists of a tolerably dark smoke grey compact limestone, and the lower part of a slightly agglutinated sand, coloured brown by carbonaceous matter. This bed, situated under the white limestone, rests immediately on a thick deposit of sandstone, and sand without shells, and may be compared to the slaty clay bed described in tke work of Messrs. Cuvier and Brongniart, page 222.

I believe that P. Lamarkii, and P. acuminatus have yet been observed only in the second freshwater formation, whilst C. lapidum has not yet been seen beyond the inferior strata of the first freshwater limestone.

The white limestone of Etampes contains, besides potamides, lymnea, planorbes, and a shell, which has as yet been but very seldom observed; it is the cyclostoma that M. Brongniart has described under the name of C. elegans antiquum. The numerous individuals that I have seen have appeared to me *always* thicker, and shorter than those of the recent species; so that I conceive they ought to be considered as a distinct species, which may be named C. crassatum. This cyclostoma has yet only been found in the second freshwater formation, whilst the C. mumia is the characteristic fossil of the first formation.

This is sufficiently evident with regard to the beds containing lymneæ, and those of the inferior white limestone, which will easily be refered to the lower part of the Paris gypsum formation. But the assertion requires explanation as it respects the green marls and mill stone without shells. Messrs. Cuvier and Brongniart, when advancing their opinion upon what they have named fresh water formations, have shewn that circumspection which always accompanies true merit; it might be said that they feared being reproached for having given too much importance to their discovery, by enlarging the extent of this new mode of formation; thus they have only attributed it to those rocks whose origin is evident, and have not hazarded any opinion respecting those in which animal remains did not exist. Yet now that the idea is more familiarized, and that these formations are known to be very abundant on the surface of the globe, and that it may be said, if the expression be permitted, that they do not cost nature more than the marine formations, now, I say, we may allow ourselves to speak with more confidence. I believe, for example, I have been able to shew* that the siliceous limestone belongs to the same mode of formation as that containing lymneæ and other freshwater shells.† This opinion affords a double motive for attributing the mill-stone without shells to the same origin; for this mill-stone is known, on the one hand, to have a great affinity to certain flints of the siliceous limestone formation, and that, on the other, it so much resembles the mill-stone with freshwater shells, that the authors of the mineral geography of the environs of Paris, were at

* Notice of the existence of freshwater limestone in the departments of the Cher, &c. Journal des Mines, tome xxxii, p. 43.

† New observations, which will be published in the second edition of our Essay on the Mineralogical Geography of the Environs of Paris, induce us entirely to admit M. d'Halloy's opinion. We have now direct proofs that the siliceous limestone forms a part of the first or lower freshwater formation; but we cannot admit the resemblance of the upper millstones, whether they contain shells or not, to the flint of the siliceous limestone. (Note of the Editor of the Annales des Mines.)

considerable trouble to find decided characters to distinguish them. It appears to me that such strong analogies, which are not contradicted by any positive facts, ought to suffice for establishing our opinion.

The same reasoning might apply in a certain degree to the green marls of the environs of Damerie, which do not contain shells, and which are placed between the two systems of fresh-water beds. I should observe, that I have never discovered traces of the marine formation posterior to the first fresh-water formation to the east of Chateau-Thierry, as if the surface of this country had been too much elevated to be covered by the waters of the sea, which have, at different times perhaps, submerged the plain of Paris; a very important fact, and one that deserves to be verified in a greater number of places.

The country, the nature of which I have just shewn, is one of the finest examples of the relations existing between geological character and agricultural productions. Corn is cultivated throughout the chalky plain; the slopes of the calcaire à cerites are covered with vines, and as this limestone, almost always friable, has fallen down upon the chalky base, the cultivation of the vine extends as far as the level of the plain; the soil of the true chalk is not generally favourable to the vine, and it is right, in order to avoid the error resulting from the habit, (when speaking of Champagne) of associating the idea of a chalky soil and a country producing good wines, to remark here, that the vineyards of Champagne are generally on the borders of this region. Those which on the western side produce the best wines, are, as has been seen, upon the calcaire à cerites formation, and those on the eastern border belong to beds beneath the chalk properly so called, beds on which I shall say a few words at the end of this memoir. When vineyards occur in the interior of Champagne, they are upon patches upon one or other of these rocks, which are found isolated on the true chalk.

The fresh-water limestone, and its corresponding marls, are not thick enough for any particular system of cultivation; often indeed the limestone beds, from their solidity, form escarpments too steep for cultivation: M. Desmarest, jun. has observed that the fresh-water marls are advantageously employed in rendering the chalky soil fit for the growth of vines. Lastly, the platforms of the millstone formation are commonly covered by forests or heaths, which, from the large blocks scattered over the surface, remind us of a primitive country.

The limits of the Parisian formations, in the part of the basin south of the Seine, do not preserve this physical demarcation characterizing those of the northern portion. This arises from the changes that have taken place in the geological structure of the surface, changes which I am about to indicate in a general manner.

Although we may be accustomed to consider the different rocks forming the basin of Paris, as placed horizontally one upon the other, and they really appear to be so in the central part of the basin, we shall observe, taking these formations collectively, that they have a southern dip, sufficiently so to represent to a certain degree, a set of wedges arranged like tiles on a roof, with this peculiarity, that the inferior wedge attains the greatest elevation.

The first series is, as is known, the calcaire à cerites, the most elevated part of which appears to be on the northern extremity of the basin among the hills of Laon, 300 metres above the level of the sea,* [984 feet] exactly where this limestone is not covered by any other formation. Quitting the summit of these hills, the level of this rock gradually becomes lower, dips under the other deposits, and disappears altogether on the south of the Marne and the Seine.

The second series, or the first fresh-water epoch, which I consider from what I have above stated, as composed of siliceous limestone, gypsum, and the first fresh-water limestones and marls, first shews itself some distance to the north of the Marne and the Seine; it is not of more than 150 metres [492 feet] elevation in the environs of Paris; but it is probable that it attains a much greater height to the

* Lemaitre, Journal des Mines, No. 35, p. 853.

eastward, especially towards the borders of Champagne, this formation occupies a considerable space of a triangular form, dips under the later formations the same as the preceding, and disappears in the neighbourhood of a line drawn from NW. to SE. which shall pass by Houdan, Arpajon, and Nemours. Throughout the greater part of this surface, that is to say, that which is covered by the siliceous limestone, the calcaire à cerites is, as Messrs. Cuvier and Brongniart have observed, altogether wanting; it is only represented in some places by the plastic clay, separating the siliceous limestone from the chalk. The formations of the third series are differently disposed from those of the two former, but before I commence any observations, I should state the geological extent to be given to it.

Messrs. Cuvier and Brongniart, with that luminous precision they have shewn throughout their work, have distinguished and characterised three separate rocks between the two fresh-water formations; namely, the marine marls above the gypsum, the sand and sandstone without shells, and the superior sand and sandstone with shells. I conceive, that considering these rocks generally, a great formation may be observed, constituting the second marine epoch of the Paris basin. The sand and sandstone without shells are, in reality, placed between two formations, whose fossil system is nearly the same. The superior is of the same nature as the rock without shells, which can only be distinguished here by this negative character; the inferior formation is not so different as it would appear to be at first sight, as it is known that the same system may be calcareous or quartzose, according to situation; this is actually the case in this instance, for at Etaples the sand and sandstone without shells rest immediately on a sandy deposit containing many shells, where a great quantity of pectunculi, cythereæ, &c. may be distinguished; that is, the same fossils as in the sand, which elsewhere covers the sandstone without shells. Finally, the absence of organised bodies in this last mass, is but the repetition of a fact, observable in numerous instances, namely, that the moluscæ diminished and even disappeared often altogether in the liquids which deposited siliceous matter.* This second marine formation thus limited, is not so *concentrated* as the first fresh-water formation. It begins to shew itself sooner northward; and generally forms but thin patches on the right bank of the Seine; it becomes thicker on the south of that river, and constitutes those escarpments, capes, and isles, which form the characteristic features of the country extending towards Fontainebleau, Versailles, Epernon, &c.; this physical character is probably owing to the facility with which water attacks this formation, the sandy mass which extended from these escarpments to the patches on the right bank being removed. I do not know if this sandy formation has been observed more than 200 metres [656 feet], above the level of the sea; in other respects, it follows the general law of a southern dip, plunges under the second fresh-water formation, and totally disappears to the south of a line drawn from Chartres to Nemours.

The second fresh-water formation, with which, from reasons before stated, I associate the millstones without shells, forms the fourth series. It covers almost the whole Paris basin; but throughout the parts where the inferior formations occur, it is often interrupted and not in sufficient force to characterize the district; advancing along the left bank of the Seine it becomes thicker, and when once the line between Chartres and Nemours is crossed, where the sandstone without shells is seen to end, it forms the only remaining Parisian formation, resting immediately on chalk, as may be observed on the edges of the basin; the deposit of fresh-water limestone is so abundant in the interior as not to allow any other formation to appear. This is certainly the most considerable deposit of this nature that has yet been described; it is also remarkable for the solidity of the stones it affords, the variety of siliceous matter it contains, and the quantity of land and fresh-water shells discovered in it.

* We entirely adopt this opinion of the author as to the extent of this second marine formation. (Note of the Editor of the Annales des Mines.)

That part of this formation which borders the Loire, and in general all that to the south of Montargis, Neuville, &c. is covered by a sandy deposit, which might to a certain degree be considered as a fifth and last series, confounding itself with the sands covering the northern part of the Sologne. This sand is generally composed of tolerably sized grains of crystalline quartz, and is accompanied by rounded fragments of the same; rounded pebbles of white flint (silex) occur on the surface, at least in the neighbourhood of the Loire. I shall return to the consideration of this sandy deposit, whose geological epoch I shall however not venture to determine; contenting myself by making known the different opinions that may be entertained on this subject, when I shall in the sequel have exposed some facts, which may serve to develope these ideas.

It will be seen, from the above, that the principal formations of the Paris basin, notwithstanding their successive superposition, have a real geographical disposition, dividing the country to a certain degree into physical regions, distinguishable by their aspect, and agricultural productions.

It may have been remarked in the first place, that the calcaire à cerites forms the greater part of the basin on the north of the Marne and the Seine, and a great band on the south of that river, between Versailles and Houdan. If this country is considered in a physical and agricultural respect, it will be observed that it has an unequal surface, furrowed by numerous valleys, of a dry nature, in which the cultivation is much varied, and remarkable for the produce of a considerable quantity of wine.

The country between the Marne and the Seine, commonly known by the name of Brie, is on the contrary wet and covered with marshes, owing to the clay accompanying the siliceous limestone, and the millstone, by which the surface is almost every where covered.

The sandy formation of the second marine epoch, has not, as a principal formation, sufficient geographical extent, to constitute a physical region; but all the platforms where it occurs, are covered by large forests.

La Beauce, or the great platform of freshwater limestone between the Seine and the Loire, is remarkable for its uniformity and the almost exclusive culture of corn.

The sandy deposit covering the freshwater limestone, forms a particular region on the SE. of la Beauce, corresponding to the country commonly named Gâtinais, which is low, wet, unproductive, and generally covered by forests.

This same sandy deposit, mixed with the mud of the Loire, gives rise, along that river from Orleans to Blois, to a vineyard so considerable, that it may be regarded as a small physical region, separating la Beauce from La Sologne.

What has been said on the physical and geographical extent of the portion of the basin on the south of the Seine, is sufficient to make known that of the limits of these countries; I shall confine myself therefore to a rapid sketch.

It has been seen in the work of Messrs. Cuvier and Brongniart, that these limits are clearly defined between Mantes and Epernon; as also to beyond Gué-de-Longroi, on the east of Chartres, where the right bank of the Voise forms a well determined inclination, forming the edges of the Parisian rocks, opposite the chalk plain of the environs of Chartres. This plain is, like those on the north of the Seine, low and even, rising towards the hills of le Perche.

These limits afterwards take a SW. direction, passing near Bonneval, and follow the banks of the Loire, at a certain distance, as far as opposite Vendome, whence they bend towards Blois. There is here no physical demarcation; the surface of the Parisian formation is almost on the same level with the chalk plain, arising from the inferior formations having ceased, as has been above stated.

It afterwards becomes extremely difficult to assign the true limits of the Parisian rocks on the south of the Loire, because they lose themselves under the sandy deposit of intermediate origin, which I have noticed above. Nevertheless some outcrops and quarries expose freshwater limestone along the left bank of the Loire, from Blois to Cosne; but with this difference, that the chalk which was hid in the northern part of the Sologne by sand, reappears beyond

Gien, and forms well defined elevations on both banks of the Loire, enclosing the small tongue of freshwater limestone along this river from Gien to Cosne; so that the Parisian formations which so long predominate over the chalk, which afterwards confound themselves with it under the same level, terminate in the end in a valley lower than the chalk hills that surround it.

If, setting out from this SE. extremity, we resume the continuation of the limits of the Paris basin, we shall see that they are not better defined along the canal of Briare than in the vicinity of the Loire, but they can afterwards be distinguished on the north of Montargis, and especially in the environs of Nemours, where the sandstones without shells appear.

Such is the extent of the Paris basin considered as a whole; for the different formations of which it is composed send off ramifications of greater or less extent, beyond the limits I have mentioned. The freshwater formation especially extends to considerable distances. I have already had occasion to remark, that it extends along the banks of the Loire and the Allier, and over the platforms of Berry.* It is known to form a new and very extensive basin in the Limagne d'Auvergne. It is found, but in small quantity, towards Tours and Mans.

The second marine formation also constitutes some isolated patches beyond the limits of the basin; at least I conceive that I can refer to this formation, the deposit of white sandstones occurring in the plains of Picardy, and as far as the departments of the North, and Jemappe, as also those existing in the chalk country between the Seine and the Loing.

Plastic clay, under the form of black pyritous earth, is also found in detached patches upon the chalk on the north of the Seine; and it appears that there are other argillaceous deposits far distant from the Paris basin, which, though resting upon older rocks, may still belong to this formation.

The true calcaire à cérites appears, on the contrary, cir-

* Journal des Mines, tome 32, p. 43.

cumscribed within the limits of the Paris basin; and this is a remarkable circumstance in the history of that formation; at least I have not observed it elsewhere, although I have made a point of examining the places, where, from mineralogical resemblances, this formation has been said to exist, and where I have recognized calcareous rocks older than true chalk, possessing the texture and yellowish colour of the Paris building stone.*

I shall conclude this memoir with a few details on the chalk formation, though, after the excellent description contained in the Mineral Geography of the Environs of Paris, there remains scarcely any thing to say on common chalk; the inferior strata of this great formation being in contact with the southern part of the Paris basin, I conceive that it will not be out of place to make its different modifications here known.

Beds more or less differing from true chalk, by their mineralogical character, by their chemical nature, and even by peculiar fossils, separate this formation from the old horizontal limestone, † but approach true chalk by insensible shades. The four following modifications may be distinguished in this series; 1st, chalk with pale flints; 2d, tuffa, or coarse chalk often chloritous; 3d, sand and sandstones of the chalk, which are almost always mixed with limestone; 4th, grey clay, generally marly, rarely plastic, and sometimes containing chlorite. The passages and alternations of these different modifications one into the other, does not allow the order of superposition to be determined in a constant manner. It may nevertheless be remarked, that the chalk with pale flints is generally the newest, and that it precedes

* This is the case with all the limestones existing beyond the chalk belt that surrounds the Paris basin, whether towards Caen, Tours, Bourges, or in Burgundy and Lorraine.

† It appears from a note to a paper inserted in the Annales des Mines for 1822, p. 367, that the author included in this "bad name" the Alpine limestone (magnesian limestone), the lias, and the oolite formation; in this case it apparently represents the latter, and perhaps also the lias. (Trans.)

the common chalk with dark flints, from which it is sometimes not sensibly distinguished; that, on the contrary, the argillaceous rocks are the first of the formation, and there is even a part of them which belongs rather to the old horizontal limestone, than to the older chalk.

Fossils are very abundant in these different systems; some, such as the echinites, are the same as those in common chalk; others, such as ammonites, resemble those in alpine limestone; there are those, such as belemnites, terebratulæ, &c. that are common to them, the chalk, and alpine limestone; those that may be cited as characteristic, as much for their abundance in these systems, as for their rarity, and even perhaps their total absence in other formations, are the orbicular gryphite, and a large shell referred to the genus spondylus.

The immense chalky basin which extends as a gulf into the north west of France, presents these different modifications of the older chalk throughout its contour, except towards the English channel, where the true chalk extends to the sea-shore. Every where else the four systems I have noticed may be recognized, with this difference, that one or two of the systems, being often considerably developed, mask the others, scarcely existing, and alone determine the character of the country. It is thus, that calcareous, sandy, and clayey countries, occur within the range of this formation. There is also this general difference, that this rock forms but a narrow band on the east, from the Oise to the Yonne, whilst on the south, and especially on the south-west, it occupies a considerable space. This appears to arise from the beds, notwithstanding their horizontal appearance, having an inclination on the eastern part proportioned to the rapidity with which the formation on which they rest, rises; hence it follows, that the same system cannot long appear on the surface. In the south-west where the surface is lower, the beds, being on the contrary more perfectly horizontal, extend over a greater surface.

The deposit of the Parisian formations is not placed exactly in the middle of the great chalk basin, for the southern portions rest upon the older chalk; but it is extremely difficult to say where the common chalk ends, since the passage of this rock into the chalk with pale flints is so gradual, that it cannot positively be determined where it takes place. I conceive nevertheless it may be admitted, that in the western part of the basin, the countries to the south-west of Chartres, Courville, Verneuil, &c. belong to the chalk with pale flints, which afterwards forms the belt of the Parisian formations to beyond the Loire.

This chalk differs but little from that with dark flints, and sometimes contains subordinate beds, which do not at all differ from it; it is generally of a coarser grain, of slighter cohesion, and contains a greater quantity of sand, sometimes clay, and even chlorite in the inferior beds; it is often advantageously employed to manure lands. The flints are generally more abundant than in common chalk; there are even situations where their mass surpasses that of the chalky matter; their colour is commonly light or yellowish brown, sometimes ash grey, rarely black. They now and then lose their mineralogical character, and pass by insensible shades into jasper, calcariferous sandstone, and breccia or puddingstone, which notwithstanding their *clastoide* appearance, clearly shew an origin analogous to that of the other siliceous nodules.

The passages of the chalk with pale flints into, and alternations with, coarse chalk, and the sands of the chalk, render it also extremely difficult to trace a limit between these two rocks. But the predominance of the sandy beds to the west of the band of chalk with pale flints, which I have just noticed, forms a sandy country, which may be considered as divided into two small regions by a point in the shape of a cape, formed by the ancient limestone in the neighbourhood of la Ferté-Bernard (Sarthe). One of these regions known by the common name of Perche, is a thicket country, furrowed by numerous small valleys, and extends from the environs of Aigle (Orne), towards Montdoubleau (Indre and Loire); the other comprehends the arid platforms between the Sarthe and Loire, and extends a little to the north of the first of these rivers.

These sands might at first sight be taken for an alluvial soil, the rather because, from their want of adherence, the higher parts have been disturbed by water, and are often mixed with rolled pebbles; but when these countries are more carefully studied, we are soon convinced that they belong to the ancient chalk formation; the chalk becomes coarser and contains more sand the deeper it gets, and the siliceous nodules often pass into the state of calcariferous sandstone. This sandy chalk is afterwards seen to alternate with regular beds of sand and sandstone, containing fossils characteristic of the ancient chalk, and the principal mass of the sandy rock can be distinctly recognized to dip under the chalk. These sands and sandstones commonly contain calcareous matter, and sometimes chlorite; there are some however altogether pure: the greater part are fine grained and uniform, others are of inequal grains; their colour is commonly yellowish, sometimes whitish, rarely blue, red, or ferruginous. This last colour belongs principally to the sandstone beds passing into puddingstone, named *roussard* by the country people, and which occur in the midst of the sands. Organized bodies are not generally found in the purely quartzose beds, but they are often very abundant in those containing calcareous matter; the most common are the orbicular gryphite, and some species of oysters. Ammonites begin to appear, or, to speak more correctly, this rock appears to be the last term of this animal's existence; it is probable however that when the species of this genus shall have been better studied, the ammonites of this epoch will be found to differ from those of the alpine limestone; the remains of fish and the impressions of vegetables are also found in it.*

* This last observation is due to the zeal of M. de Maulny, naturalist of Mans.

I ought to remark before I quit this country, that I do not consider that all the sand covering the platforms between the Sarthe and the Loire, belongs exclusively to the chalk formation; for the presence of thin mill-

La Touraine occurs to the south of this sandy country, a region that extends to the country of the ancient horizontal limestone, which is met with to the south of Chatellerault, and Chatillon sur Indre, and whose surface is formed of coarse chalk. This substance, known in our western departments by the name of *triffeau* is sometimes tender and friable; at others, hard enough to form good building stones; its colour is most commonly yellowish white, often having a greenish tint, arising from the presence of chlorite; the flints in it are almost always whitish, often passing into the hornstone variety, sometimes into jasper and calcariferous sandstone. Its fossils are very abundant and extremely various: the orbicular gryphite is particularly distinguishable. The thickness that the beds of tuffa ordinarily possess, the facility with which they are worked, the double advantage derived from them for building, and the manuring of ground, have given occasion to the hollowing out of immense quarries, sometimes inhabited by modern troglodytes. These quarries form one of the principal features of this rock, which are again found in the tuffas of the Meuse Inferieure, as well as in those of the banks of the Loire.

The tuffa of Touraine is covered by a thick bed of sand, full of whitish flints, and sometimes mixed with clay, which is only the sandy chalk washed by water; to these two systems is owing the contrast presented in the agricultural condition of this country. When the surface is cut into sufficiently deep to expose the tuffa bed, it becomes extremely fertile, and merits the epithet of Garden of France that has been given it. But the platforms covered with sand and flint are absolutely arid, and only afford extensive heaths.*

stone beds scattered over some of these plateaux, joined to the existence of a small deposit of freshwater limestone near Mans, renders it probable that there exist some superficial deposits of freshwater sands; but I have not been able to verify this fact in a positive manner.

* The marl, or decayed shells (falun), found on some of these platforms, contributing, as is known, to fertilize them, is a distinct deposit much more modern than the tuffa. The shells composing it, of which M. de Tristan, an able naturalist, is preparing a description, possess

La Sologne is known to be a low and marshy region, of little fertility, and of a sandy nature, situated to the south of the Loire, and to the east of la Touraine; its southern part belongs clearly to the chalk formation; the same sands mixed with unrolled flints may there be recognized as in Touraine. There is only this difference, that the surface is less laid open, consequently the tuffa is more rarely exposed, and lastly, this tuffa is not so well characterized, and more approaches the marly chalk.

But the part of this country on the north of la Saudre, is covered by a sandy deposit, the origin of which is not so easy to determine. This sand is the same as that of which I have spoken above, as covering the freshwater limestone of le Gâtinais, that is to say, it is composed of grains of white quartz commonly rounded or globular, often very large, sometimes extremely small; it is accompanied by fragments of transparent quartz, commonly white, rarely greyish, and by yellowish-brown flint, all more or less rounded, and apparently only found on the surface.

An alluvial origin has often been attributed to these sands; but according to this hypothesis, the debris of the various rocks, of which the neighbouring country is composed, ought to be found there, as actually takes place in the true alluvion of the Loire, where the mica and felspar of the Auvergne granites may very easily be recognized, even in the finest sand. There does not exist a country so exclusively quartzose, that the destruction of its rocks should give birth to the sands in question; the supposition of such a country entirely destroyed or hid, is much more contrary to what we know of nature, than the opinion admitting these sands to have been formed such as they are, in the same

many relations in common with the Calcaire à cerites of Paris. But the *falun* of Touraine differs from this last formation, in not passing into a stony state, and in only affording, as Réaumur has already observed, the remains of shells more or less broken.

More or less extended patches of the freshwater formation, are found on the surface of this country, either in the state of shelly limestone, or that of siliceous limestone.

manner as the other sandy rocks, whose local formation is now well demonstrated, as well by alternations with other rocks, as by the fossils they contain.

The first idea that this hypothesis presents, is to consider the sands of the northern Sologne as belonging to the ancient chalk formation, as also those of the southern part of the same region, of la Touraine, of le Perche, &c. The existence, in these last, of large grained beds, resembling the sand between the Loire and the Saudre, supports this opinion. But, on the other hand, the presence of these sands on the freshwater limestone of the borders of the Loire and the Gatinais, that of small patches of analogous sands, on the same limestone, in other places nearer Paris, as at Etampes, at Rambouillet, &c.; lastly, certain relations that they bear to the millstone formation, might give rise to the idea of their being the last term of the second freshwater formation of the Paris basin, such as Messrs. Cuvier and Brongniart have considered the sands to be that occur on the top of the hills of Longjumeau.*

I confess that I am still at a loss to decide between these two opinions, and if I had not comprehended the countries between the Loire and the Saudre in the Paris basin, I had determined, for want of geological reasons, from the consideration of physical geography alone, not to dismember so natural a region as la Sologne. It is proper nevertheless to remark in this respect, that in the hypothesis that all the sands of this country belong to the ancient chalk, it is very easy to conceive their extension over the freshwater formation; for this deposit of moveable matters, situated precisely at the opening of the great water courses descending from the mountains of Auvergne, ought to be more disturbed by the waters than those which occur under different circumstances; and some grand catastrophes, such, for instance, as that which has overwhelmed the animals of the marl and gravel, might be sufficient to throw a part of these sands over on the little elevated edge of the freshwater limestone

* Min. Geo. of the Environs of Paris, p. 55.

country, where they are now observable. It is to causes of this nature that the presence of the rolled pebbles must be attributed, which occur more or less superficially in these sands.

The chalky belt of the Paris basin is, it may be said, broken to the east of la Sologne by the point formed of fresh-water limestone along the Loire to Cosne, where it approaches the ancient horizontal limestone; but is again found beyond this point where the last systems of the formation constitute a small physical region, covered with trees, hedges, and meadows, known by the common name of Puysaie, which extends from the valley of the Loire to that of the Yonne, embracing the greatest part of the country comprehended between Cosne, Montargis, and Auxerre.

The surface of this district, less even than that of Sologne, more frequently exposes the different ancient chalk systems, such as the chalk with pale flints, the sand, and above all the clay, which is most abundant, and which gives a character to the region. In the series of these deposits one occurs very remarkable for its economical utility; it is the Pourrain ochre, which occurs in the midst of irregular beds, and more or less mixed with sand, clay, marl, and even calcareous matter, in which may very clearly be seen the series of insensible shades that mineralogically unite the quartzose substances designated by the names of flint, jasper, and sandstone.

The clay and sandy beds of la Puysaie cease nearly in the direction of a line from Châtillon sur Loing to Joigny; there then only remains the chalk with pale flints, which, on the north of Montargis and Joigny, tends to approach the true chalk, which occurs extremely well characterized in the plains of Champagne to the north of the Yonne. The space occupied by the ancient chalk afterwards narrows considerably, and forms, as I have already noticed, but a narrow band extending the length of Champagne from the Yonne to the Oise. This band, already described by M. Desmarest, is remarkable for its continuity for so great a length, and the uniformity with which it presents itself under the form of a clayey valley, bounded on one side by chalk platforms, and on the other by those of ancient horizontal limestone; for it is to be observed that the rock of Champagne, which formed but a low plain when it appeared from under the Parisian hills, gradually rises, attains a height that appears at least equal to that of those hills, and terminates at the eastern edge of the region, by a kind of escarpment which exposes the marly clay beneath the clalk; this same clay rests upon ancient horizontal limestone, which soon rising to a level above that of the chalky platforms, prevents the extension of the clay. It also appears, that the property this system possesses of being easily worn by water, has considerably contributed to the valley that it now presents, the more so as in the places where this kind of valley is traversed more or less transversely by a river course, it forms large swellings out, the argillaceous soil of which is mixed with a great number of small rolled pebbles, of a calcareous nature. There occurs among others one of these swellings out at the place where the Marne and its branches traverses it, known by the name of Perthois, and which is remarkable for its great fertility.

Although the argillaceous rock principally characterizes this border of Champagne, the other systems of the ancient chalk are not altogether wanting; even the tuffa with chlorite occurs, especially at Autry, in the department of the Ardennes; but scarcely any flint is there seen, and it is a very remarkable fact that the ancient chalk of Champagne differs from that of the other parts of the basin, by the same character that is peculiar to the true chalk of the same district.

The limits of the chalk formation are too far removed from the Paris basin, on the north of Champagne, to be considered in this memoir; but there is very near this basin, and even at a short distance from Paris, a very small district, where not only the ancient chalk is seen to appear, but also the formation or at least the last member of the limestone formations older than the chalk. This canton, commonly called the Pays de Bray, is situated at the confines of the departments of the Oise, the Seine-Inferieure, and the Eure. It resembles an island that may be considered as the summit of a mountain buried under the great chalky deposit.

The parts of this deposit in the neighbourhood of the Pays de Bray are remarked, in the first place, to acquire the characters of ancient chalk; between Argueil and St. Sausom (Seine Inferieure) that substance is seen which is filled with a great quantity of blackish green chlorite grains, and another modification of a coarse texture, passing into the state of sandy marl, and containing nodules of greyish calcariferous sandstone instead of true flints. It afterwards appears that the sands and marly clay, forming the peculiar character of the country, rise from beneath this coarse chalk; I only say appears, because the moveable nature of these deposits, and the labours of agriculture, conceal the superpositions, and that, on the other hand, the neighbourhood of the sands and plastic clay of the Parisian calcaire à cerites, would allow the supposition that that formation had extended as far as the Pays de Bray. But the presence of limestone that occurs in the central part, among other places, at Meuerval, Cuy-St.-Fiacre, &c., does not leave a doubt, that the greatest part at least of the clays of this canton belong to the formation intermediate between the chalk and horizontal limestone. This limestone, commonly yellowish-white, or yellowish grey, is remarkable for its hardness, the abundance of spathose parts it contains, and above all for the great quantity of small oysters that enter into its composition, although there are neverthcless some beds quite compact, and without fossils. One cannot very well judge of the position of its principal mass with regard to that of the clay; but it is clearly seen that beds of the two systems alternate with each other.

These sketches suffice for recognizing a small formation, very remarkable for the constancy with which it presents the same mineralogical and geological characters in very distant countries, as le Berry, Lorraine, the Boulonais,* the coast of Calvados, &c. This limestone is every where distinguished by its tenacity, by its spathose parts, or by a texture which, without being spathose, nearly approaches the crystalline state, by the abundance of its fossils, which afford besides the oysters of the Pays de Bray, other species of a considerable size of the family of the ostracæ and of that of byssiferi, trigoniæ, a great quantity of zoophytes, particularly madrepores, &c. This limestone is always in the neighbourhood or accompanied by the clay below the chalk; its geological position is no where better seen than at the cliff named Vaches Noires, on the coast between Honfleur and Dives (Calvados). It there forms some beds more or less thick, situated between two systems of grey marly clays. The superior system often contains blackish green grains of chlorite, and passes into the chalk with chlorite, which is immediately above it. The inferior clay is characterized by great and wide gryphites (gryphæa latissima),† and rests

* The major part of the Boulonais consists of the same formations as the Pays de Bray; it is only in the northern portion that from beneath these rocks rise successively, the coarse horizontal limestone, then the marquise marbles, which I consider as belonging to the older alpine limestone, or zechstein of the Germans, and lastly the coal measures. The latter dips again under the chalk, which borders this little region by a chain of hills in the form of a semi-circle. It is right to remark here in consequence, that I committed a mistake, in 1808, (Journal des Mines, tome 24, p. 348) by referring the Boulonais limestone, which I had not myself seen, to the transition formations, abundant in the north-west of France.

† Coral rag, which rises from under the superior rocks, at Henqueville cliff (between the Vaches Noires and Honfleur), and forms the top of the hill on the east of Benerville, fines off on the eastern part of the Vaches Noires; the following is a section of this cliff a little to the west of Oberville, beginning with the surface.

Green sand formation
Limestone (inferior to coral rag)
Whitish marl
Thick oolite bed
Oxford clay.

The gryphæa latissima is, I suppose, the gryphæa dilatata of Sowerby, which I observed to be most abundant there; for further particulars and upon the oolite limestone, having a slaty structure in its higher parts, that extend towards Caen.†

sections of this coast see my paper in the 1st volume of the Geological Transactions, 2nd series. The limestone mentioned by M. d'Halloy would appear to belong to the coral rag series, and therefore is not necessarily connected with the clay of the green sand formation, though they may frequently come together, the intervening beds being wanting. (Trans.)

† The upper slaty part is forest marble; the Caen freestones occupy the same geological position as the Bath or great oolite. (Trans.)

Extract of a Memoir on the possibility of causing fresh-water moluscæ to live in salt-water, and marine moluscæ in fresh-water, with geological applications. By M. BEUDANT.

Annales des Mines, 1816.

Read at the Royal Academy of Sciences, May 13, 1816.

THE rocks of the country round Paris are known to be divided into marine and fresh-water formations; this distinction has been made because the shells found in the former are only analogous with those that exist in the sea, while the shells found in the latter are analogous with those that live in fresh-water.

M. Beudant had however discovered (in 1808), in the sandstone of Beauchamp, near Pierrelaye, a mixture of marine and fresh-water shells; he has since observed the same circumstance in a marl bed in the environs of Vaucluse. There are, besides, many other shelly beds, about the origin of which naturalists are divided, not being yet agreed with respect to the analogy of a part of their fossil shells: such, among others, is the shelly bed observable in the neighbourhood of Mayence.

These considerations, joined with that of the perfect preservation of the shells, rendering it reasonable to believe that they have not been transported, but that the animals which inhabited them have lived in the same places where their remains are now found, caused Mr. Beudant to conceive it possible for moluscæ, naturally marine, to live in fresh-water with fresh-water moluscæ; or, on the other hand, that fresh-water moluscæ could live in salt-water, with marine moluscæ.

This hypothesis received some support from facts long known to naturalists. Many species of marine moluscæ, especially oysters, cerithia, and common muscles, are known to live at the mouths of rivers, and even at distances from the sea, where the water is fresh, or at least but rarely salt; many marine fish are also known to mount rivers to a still more considerable distance from the sea, where they can never be under the influence of salt-water.

The contrary, i. e. the presence of fresh-water moluscæ, or animals in the sea, does not appear to have been observed; nothing is known in this respect but some vague notices of fresh-water fish living in the waters of the Baltic.

It was of consequence to verify this double hypothesis in a more precise manner; with this view M. Beudant undertook many series of experiments, an account of which he had the honour of presenting to the Academy of Sciences. The object was: 1. to endeavour to make fresh-water moluscæ live in salt-water: 2. to habituate marine moluscæ to live in fresh-water. Two other subsidiary objects were also proposed: 3. to search for the cause of the almost total absence of fossil shells in the gypsum beds: and 4. the cause of the moluscæ and other organized bodies, living in the Asphaltic lake, whose waters contain, according to Lavoisier, 0,44 of saline matter, of which there is only 0,06 of muriate of soda.

He commenced his experiments on fresh-water moluscæ at Paris, in 1808 and 1809: and it was not until 1813 that he could execute those on marine moluscæ at Marseilles.

He employed in the whole many hundred individuals, of which he had previously determined the species; he kept a regular journal of his observations, and especially of the number and species of the individuals that perished at different times.

Being obliged to keep these moluscæ in vases, where they were necessarily much confined, and to nourish them with food, probably not very proper for them; he was sensible that these inconveniencies were the cause of a mortality the results of which ought to be subtracted from the general total, in order to become exactly acquainted with the number of those which perished from the change in the nature of the water.

In consequence, he always took care to divide the moluscæ intended for his experiments, into series, identical as to the species and number of the individuals. One of these series was kept in water proper for it, the other was placed in a different water; thus when he experimented on fresh-water moluscæ, he kept half of these animals in fresh-water, frequently renewed, and put the other half into water made salter by degrees. He kept an account of the mortality in both cases, and he could consequently judge from the different results, of the number of individuals that perished from the saltness of the water. An inverse precaution was taken when he experimented on marine moluscæ.

The following are the principal results obtained from his experiments:

1. Fresh-water moluscæ perish immediately when plunged *suddenly* into water as salt as our seas.

Marine moluscæ do the same when *suddenly* plunged into fresh-water.

2. Many fresh-water moluscæ, could, in a very short time, be made by degrees to live in water which was gradually salted to the ordinary saltness of the sea.*

Many marine moluscæ could also, by gradually diminishing the saltness of the water, be accustomed to live in fresh-water.†

* Among the fresh-water univalves, there are species, of which, in five months and a half, 57 individuals only perished out of 100 with salt water; and 54 out of 100 died in the same time kept in fresh-water; (all the lymneæ, the planorbes, the physa fontinalis, the ancylla fluviatilis, cyclostoma obtusa, paludine porte-plumet). The difference therefore was nearly nothing; it was considerable with other univalves, and the bivalves were not able to support the change in water as salt as the sea, (see article 3.)

† Only 36 to 37 individuals in 100 perished in five months, whilst out

3. This power of accustoming themselves to live in a liquid of a very different degree of saltness from that in which they usually exist, is not the same in all moluscæ.

Thus, among the fresh-water moluscæ, the bivalves (anodontæ, unio, and cyclas) could not live in water that had acquired the saltness of the sea—(0.04.)

Among the marine moluscæ, the patellæ, the fissurellæ, the crepidulæ, the pectens, the limæ, &c. could not accustom themselves to live in fresh-water.

4. All moluscæ, whether marine or fresh-water, can easily be made to live in water about half as salt at that of the sea, that is, containing only two parts of muriate of soda in a hundred of salt-water.

5. Fresh-water moluscæ could not in any manner be accustomed to live in water charged with sulphate of lime.

The author presumes that it is the same with marine moluscæ; he has not however put them to this test.

6. Marine moluscæ can live in waters much more charged with salt than those of the sea usually are, even when the water is saturated with it; they perish however when the liquid is over saturated, and a deposit of salt commences. M. Beudant made some experiments with water charged with carbonic acid, mineral acids in small quantities, or with 0.02 of sulphate of iron; he plunged fresh-water moluscæ suddenly into it, and they all died immediately.

The author acknowledges that it would be desirable to repeat and execute his experiments on a greater scale; yet he remarks that he has tried a number of individuals and a great variety of species, and he observes, that having experimented with vases of a small capacity, and generally under circumstances not very favourable to the change of element to which he would submit the moluscæ, it is more than pro-

of an equal number kept in salt-water frequently renewed, 34 in 100 died in the same time.

The author finished by placing, at the end of summer, the marine moluscæ, thus habituated to fresh-water, into a piece of water in a garden, where they were still living in company with lymneæ and other fresh-water moluscæ, at the end of the following April.

bable that there would be more complete success if the operations were carried on upon a greater scale; that consequently, in the changes and transitions of this kind, that are supposed to have taken place in nature, the moluscæ may be presumed to have been better able to resist them, always finding their proper food, and not suffering the constraints of every kind which must affect them in our small apparatus.

Supporting himself then by these considerations, and applying the results of his experiments to many known geological facts, M. Beudant believes himself able to draw the following conclusions:

1st. Since the same water, whether fresh or as salt as our seas, or what is better, brackish, can at the same time support marine and fresh-water moluscæ, it may be presumed that similar circumstances have existed in nature, and that it is to these circumstances we owe the presence of marine and fresh-water shells in the same bed, admitting, what every thing seems to prove, that the shells are found in the places where they lived.

2d. It may even be conjectured that in the interval between the existence of fresh and salt water in the same place, that the water was brackish, and supported at the same time animals peculiar to both; that consequently, between the beds formed by salt water, containing only marine shells, and those formed in fresh-water, containing solely fresh-water shells, we ought to meet with other beds formed of the passage of one into the other, containing marine and fresh-water shells at the same time.*

3d. If we could suppose against all appearances, with some naturalists, that the rocks named fresh-water rocks had all been formed under the sea water, the otherwise singular absence in the beds of fresh-water bivalves, of the genera, anodonta, unio, and cyclas, might be explained by the above experiments.

* M. Beudant refers the sandstone of Beauchamp and the marls of Vaucluse to an analogous circumstance.

4. It can be easily conceived why the gypsum masses of all formations, though frequently subordinate to rocks full of shells, should never contain any, since fresh-water moluscæ cannot exist in water charged with sulphate of lime.

5. The absence of living organized bodies, in the asphaltic lake, if it be really true, cannot be attributed to the great proportion of muriate of soda, since marine moluscæ can live in water saturated with it;—the cause ought rather to be sought in the presence of the bitter muriates of lime and magnesia, which are there in much greater abundance, also perhaps, in that of bituminous substances.

6. It is not surprising that masses of rock salt never contain fossil shells, since marine moluscæ perish in salt-water as soon as it is over saturated with muriate of soda.*

7. Lastly, if it is admitted that marine and fresh-water moluscæ can, under certain circumstances, live in the same liquid, inhabiting fresh or salt-water does not appear to be a sufficient motive for establishing particular genera; these distinctions of genera can only be founded on essential and constant difference in the shells, or, which is better, in the animals that inhabit them, when they can be observed.

* The author could have added another conclusion, from rock salt being always or nearly always accompanied by gypsum; from which it cannot be doubted that the water that deposited the muriate of soda, was charged with sulphate of lime. (Note of the editor of the Annales des Mines.)

On Gabbro (Euphotide of Hauy). By M. Von Buch. Extracted by M. de Bonnard.*

(Annales des Mines for 1816.)

THE collection published under the name of Magazin der Gesellschaft natur-forschender Freunde, zu Berlin, contains (t. IV. Berlin, 1810) an interesting memoir by M. Leopold von Buch, upon the rock formed of diallage, united with either jade or felspar, or with both these substances, a rock to which he assigns the name of Gabbro, given it by the Florentines. In the vii volume of the same collection (Berlin, 1815) M. von Buch has inserted a supplementary notice to his first memoir.

The gabbro, spread over the four quarters of the world, forming extensive rocks and entire mountains of several thousand feet of elevation, has, until now, almost always been misunderstood and confounded with other rocks, under the names of granite, serpentine granite, serpentine rock, and grünstein.

Saussure was the first to make known the great quantity of blocks, formed of jade and smaragdite, found on the mountains of the Jura and the hills of the Pays de Vaud. He was the first also to describe and class the two substances of which this rock is composed, as distinct minerals. Since then, M. Haüy has united jade and felspar. Some mineralogists have thought that smaragdite or diallage ought also

* Diallage Rock. (Trans.)

to be united with hornblende (amphibole), others wish to separate green diallage from schiller spar (diallage metalloïde), and to form two species of them. M. von Buch conceives, from the exterior characters of jade, and above all from the chemical analysis made of it by M. Theodore de Saussure, and repeated by Klaproth, that it ought not to be united to felspar, so long as it is not found crystallized in forms, which shew this union to be necessary. He believes also that the cleavage, and all other observations already made, tend to unite schiller spar and green diallage, and to separate the first from hornblende, and the second from actynolite (stralstein); thus it is preferable to preserve the two species, such as they are determined by Saussure.

Although blocks of gabbro are spread over the Pays de Vaud and the environs of Geneva, the position of this rock was not known in the Swiss Alps. In 1807, M. Struve and M. von Buch found in the valley of Saas, near the village of the same name (in the Haut Valais), an enormous quantity of these blocks, brought down by all the glaciers, which descend from the neighbourhood of Mont Rosa; ascending towards the glacier of Mont More, by the road of Macugnaga, in the valley of Anzasca, they met with gabbro in place before they reached the glacier. The jade is greyish white, exactly resembling that of the Pays de Vaud, and the diallage of a beautiful green colour, in pieces which are often half a foot long; the rock contains also small plates of talc, actynolite in little radiated bundles, and red garnets. The gabbro rests on mica slate, and appears to form the summit of the ridge descending from Mont Rosa, and separating the valley of Saas from that of St. Nicolas, nearly as far as Stalden, where it constitutes an enormous cap: this rock is several thousand feet in height, and about two or three German miles in length. No serpentine is found with the gabbro in the valley of Saas, but it occurs in that of St. Nicolas. The heights of Mont-Cervin, and the pyramid of Breithorn, are known from Saussure, to be composed of serpentine. In the Grison mountains, a chain descending from Mont Cimult, or Salamont, and separating the valley of Julier, from that of Err, appears to be principally formed of serpentine and gabbro, resting on primitive schist, and constituting a well characterised and extensive formation.

Gabbro was known to artists, long before the time of Saussure. The Grand Duke Ferdinand de Medicis, caused numerous blocks of it to be transported from Corsica to Florence in 1604, where it received the name of Verde di Corsica, and where, under various forms, it ornaments the Laurentine Chapel; but none has been since brought from thence, and the precise place of its occurrence was forgotten, and it was not till within these few years, that Messrs. Muthuon and Rampasse, French engineers, again found it, first in large blocks near the village of Stazzona, to the north-west of Corte, and afterwards, in place, but always accompanied by serpentine, in the high mountains of San Pietro di Restino, which form a chain between Corte and the sea. Thus there, as in Switzerland, gabbro forms a rock of itself, and not a bed subordinate to another. In the suite of rocks sent by M. Muthuon to the collection of the Administration of Mines, the most insensible passage of green diallage into schiller spar, is observable in the specimens of gabbro.

Saussure describes (§ 1313), a gabbro at the mountain of Musinet, near Turin, the mode of occurrence of which he has not observed, formed of white or lilack jade, and of diallage sometimes green, at others grey; M. Berger notices diallage mixed with the semi opal of Werner (silex résinite) at Baldassero, near Ivrée: the native magnesia of Giobert occurs there in small beds.

A part of the mountains surrounding Briancon and la Grave in Dauphiné, appears to be formed of gabbro; serpentine and talcose rocks are known to occur in the same mountains, which also furnish the substance called Briancon chalk.

The gabbro formed of jade and green diallage does not appear to have been known to the ancients. That formed of jade, felspar, and grey diallage, has, on the contrary frequently been worked by them. Antique columns formed of this rock, have, in modern times, been transformed into vases, which ornament the Vatican Museum, and it is probable that the masses from which these columns were taken, came from Egypt, as did the ancient granites and porphyries. M. Werner possesses in his collection pieces of schiller spar, (diallage metalloide) with felspar, found by Mr. Hawkins near Famagusta, in Cyprus, who states that the celebrated copper mines worked by the ancients at Cyprus, were in a rock of this nature.

Gabbro is common in Tuscany. Targioni Tozzetti classes and describes several varieties of it, which he names Nero di Prato, Verde di Prato, Granito dell' Impruneta, and Granito di Gabbro. The two first varieties are rocks having a serpentine base, mixed with schiller spar (diallage metalloïde); the two others are formed of yellowish white felspar, and of greenish grey schiller spar with a little jade. A fifth variety is a serpentine rock resembling those of Saxony. Targioni mentions numerous situations where these rocks are found in the environs of Florence and Leghorn; but neither the mountains of Prato, nor those of Leghorn, have been visited by modern geologists. The old observations of Targioni shew how much our gabbro is united to serpentine, and how much they agree in position, since he calls them both by the name of gabbro.

The rocks of Covigliano and of Pietramala are formed of gabbro, the principal mass of which is composed of diallage in small grains, of a leek green colour; it contains grey felspar in small grains, a great quantity of white specks, which appear to be steatite, small bundles of actynolite, brilliant points of pyrites, many small veins filled with calcareous spar, and geodes covered in the interior by transparent double-pointed rock crystals; black masses are also observable, apparently of serpentine, which seem clearly to shew the passage of gabbro into that rock.

Gabbro occurs also abundantly in the environs of Genoa; the high mountains separating the gulf of Spezzia from Mont Ferrat, appear to be almost entirely composed of it. Dr.

Viviani observed it in 1806, and described it in the Journal de Physique. Its mass is composed of white felspar and jade; diallage occurs in blackish green fragments, of a very laminated texture, possessing a metallic lustre. Dr. Viviani has observed the relations that exist between serpentine and gabbro, which he proposes to name serpentine granite. M. von Buch had found it in another part of the same mountains in 1799, where the rocks become more ancient, as they recede from the sea. Quitting Spezzia a fine grained grey-wacke is first observed, then a red and black transition limestone; nothing but gabbro is met with from Borghetto to Matanara. Near Sestri it is covered by schist worked as roofing slate at Lavagna and Chiavari. Lastly, Saussure mentions (§ 1362) a species of granite composed of white jade, a little granulated, and of laminated grey diallage (smaragdite) which he found near the castle of Inerca, in the Riviera de Ponenta, succeeding suddenly to the serpentine and other talcose rocks, which cover mica slate near Voltri.*

* M. von Buch observes that the primitive rocks of this country are exposed to view on the sea coast only, near Voltri, at Savona, and at Cape Nolis; but that no more traces of them are observable, either towards Nice, or in the interior of the country, in the direction of the great mountains of Piedmont. The opinion of those then, he adds, is erroneous, who judging only from geographical maps, make the Corsican mountains a continuation of those of the environs of Genoa; they unite things that are entirely different. It would be as extraordinary to affirm that Cape Corte, formed of granite, is a continuation of Cape Delle Melle, composed of black transition limestone, as it would be to affirm that the Vosges is a continuation of the Jura.

We shall remark, upon this head, that a similar error is committed every day by those who make maps, and by those, who, reasoning from them believe that the mountains of Hunsdruck and Eiffel, on the left bank of the Rhine, are the continuation of the Vosges. The chain of the Vosges, with a direction from south to north, formed almost exclusively of sandstone from the latitude of Strasburg, and diminishing in height in its northern parts, finishes a few leagues south of Mont Tonnere, in the hills and plains of the Palatinat. The schistose mountains of Hunsdrück, on the contrary, unite themselves with those of the same nature of the Ardennes, and form a chain, having a direction from S.W. to N.E., separated from the Palatinat by the porphyries of Mont Tonnere, of Greutznach, &c. and by the trap formation of the banks of the Nahe.

M. von Buch observes that the geological position of gabbro is that always assigned by the German geologists to serpentine, and that these rocks are almost always found near each other, and even mixed; it appears to him probable that serpentine is but gabbro mixed with a great quantity of talc, and in which the constituent parts are, from their very great fineness, no longer distinguishable. He observes, in support of this opinion, that serpentine is not a simple mineral, and that it is an error to assign it a place in mineralogical systems, that it contains parts differing from each other in colour, fracture, weight, and hardness, and which, where they acquire a visible size, appear as diallage, talc, oxidulated iron, mica, &c.: lastly, that Rose has remarked oxide of chrome as a constitutent part of serpentine, this M. Vauquelin has also found in green diallage.

The Zobtenberg, in Silesia, described long since as a mountain of serpentine, is entirely formed of gabbro resting upon the serpentine worked at the foot of the mountain, near the town of Zobten. The same rock is found in many places in the county of Glatz, and in the mountain of Herthe, near Frankenberg; it is composed of white felspar and grey diallage with a little jade, and contains pyrites. M. von Buch confesses that at the time of his travels in Silesia, in 1796 and 1797, he considered this rock as formed of felspar and hornblende, though the want of a double cleavage made him even then doubt the correctness of this idea; and that he had in consequence classed it among the primitive greenstones (diabase, grünstein). Shortly afterwards, the identity of this rock with those of Prato and Genoa being admitted, M. Karsten classed these last also with the greenstones (grünstein) in his Mineralogische Tabellen; but primitive greenstone, says M. von Buch, is composed of felspar and hornblende; whereas gabbro is formed of diallage, jade, and felspar. The geological position besides of these two rocks is not the same; confusion will therefore arise if they are called by the same name.

The town of Vienna is entirely paved with gabbro. The old sandstone and limestone pavements resisted but a few years the causes of destruction which the movement of a large town brings with it; the great hardness and peculiar tenacity of gabbro, renders this pavement at present very superior to that of all other capitals. It is quarried at Langenlois, near Crems, in Lower Austria; there are no data upon the geology of this country, but it is known that a great quantity of serpentine rocks, mixed with different minerals, and, even as it appears, with green diallage, occurs on the left bank of the Danube, near Gottweig, and not far from Crems.*

M. Esmark had observed, in 1802, that the mountains of Thron, situated in the eastern part of Norway, between Rorass and Fordel, elevated 4000 feet above the sea, were formed of a peculiar rock, composed of felspar and a substance *which resembled hornblende without being hornblende.*

In 1806, M. von Buch found gabbro on the western side of Norway, three miles to the south of Bergen. It is composed of felspar, and grey diallage, without jade. It constitutes the whole of a branch of mountains, extending for several leagues on the right side of Saumangerfiord, and often forming precipitous rocks. The mixture is never large grained; the grain is sometimes so fine that the diallage can with difficulty be distinguished; but elsewhere, and more particularly on the side of a lake near Kallandseid, the crystals of diallage are often the size of the fist, their greenish grey colour, and the lustre they exhibit on a single large surface, easily distinguish them from hornblende, beds of which occur near in argillaceous slate. Gabbro reposes here on primitive argillaceous slate that rests on gneiss.

The same relative position occurs in the neighbourhood of the North Cape, in the isle of Mageroë. The rocks are there so much exposed, that the changes that take place in their nature are easily observed: and a complete geological passage may be traced from the primitive schist to large grained gabbro. The nearest schistose rocks that surround

* Stütz oryctographie von Nieder Œsterreich, p. 228.

Kielvig almost resemble mica slate; they dip rapidly towards the N.W., that is, towards the interior of the island, and thus serve as a base to the other rocks of which the island is composed. A fine grained granite is observed upon the heights, resting upon the schist,—this granite contains some scattered plates of black mica, with an abundance of hornblende. Diallage shortly appears as a constituent part of the granite, and in a few more distant rocks the granite is changed into a fine grained gabbro. Further still, towards the centre of the island, gabbro forms masses of 1400 feet in height. It is very large grained, and altogether resembles that of Zobtenberg and Prato. The grey diallage has a laminated structure and is very shining, and a little conchoidal in the cross fracture; the crystals are often found almost perfect, and in the form of a four-sided prism terminated by a four-sided pyramid. The diallage resists decomposition better than the felspar, and forms salient crystals on the surface of the blocks. The extreme rocks of the North Cape are formed of gneiss in very thin laminæ, apparently subordinate to mica slate, the rock generally observable on these coasts.

It will be observed that gabbro rests on primitive schist in the North. We have seen it at Genoa under transition slate; its geological place appears therefore to be well determined.

It seems also that gabbro occurs with serpentine in Norway, which accompanies it in other situations. This circumstance appears to M. von Buch to be easily explained; he considers serpentine, as we have seen, to be a mixture of different minerals undeterminable from their small size.

M. de Humboldt observed gabbro near Guancavelica, above the Havannah, in the interior of Cuba, as also widely extended masses of serpentine containing a great quantity of schiller spar (Diallage metalloïde).

Gabbro ought therefore to be considered as a rock widely spread over the surface of the earth, which follows next, in the series of formations, to primitive schist, and is anterior to porphyry. The gabbro formation is intimately connected (geologically considered) with the serpentine formation, and the last is in general anterior to the former.

When extracting from M. von Buch's memoir, we have thought it right to preserve the name given by him (after the Florentines) to the rock of which he treats; but we repeat, this rock is the one described under the name of Euphotide, by M. Haüy in his lectures, and by M. Bronguiart in his essay on the classification of rocks.* The term gabbro has moreover been applied to "amphibole hornblende."† That of Saussurite, which has been proposed for jade, and which could scarcely be adopted for that substance, whose present name is so universally spread, would perhaps have been perfectly suitable, applied to a rock hitherto undescribed, and to which de Saussure had just called the attention of mineralogists. Gabbro or euphotide exists without doubt in many other countries than those where M. von Buch has observed it. It appears to us that it has, as in the case of Silesia, often been described by the German mineralogists under the name of primitive greenstone (grünstein), many of whom, as we have before observed, believe that diallage and hornblende ought to be considered as a single species. We shall cite as examples the greenstones of Harzeburger forst, of Baste, of the valley of Radau, &c. in the N.E. part of the Hartz. These rocks are certainly, or at least in part, euphotides. It is remarkable that they are associated with serpentine, containing schiller spar (diallage metalloïde) so well known under the name of schillerspath or schillerstein of the Hartz.

Euphotide and serpentine have been observed at the western extremity of Cornwall; the Lizard point is formed of it. Observations on this subject are contained in Dr. Berger's memoir inserted in the first volume of the Transactions of the Geological Society of London.*

* Journal des Mines, No. 199.

† Memoir on Basalt. Journal de Physique, 1787.

* Professor Sedgewick has published a detailed account of the diallage and other rocks of the Lizard district in the 1st volume of the Cambridge Philosophical Transactions. A description and coloured map of the same district is given by Mr. Majendie in the 1st volume of the Transac-

tions of the Geological Society of Cornwall, and in the 2nd volume of that work, there are some observations by Mr. Rogers on the same place.

Dr. Mac Culloch mentions diallage rock (Classification of Rocks, page 645, &c.) as abounding in the islands of Unst, Balta, and Fetlar, and as occurring also, but in very small quantity, in the northern extremity of the Mainland of Shetland. In Unst and Fetlar it is in contact with serpentine. The diallage rock is described as stratified, and alternating with primary schistose rocks, and serpentine.

Dr. Mac Culloch, in his synopsis of this rock, (p. 649, &c.) separates it into three divisions, as follows.

First division.

Simple; or of diallage alone.

A. A confused mixture of crystals of diallage.

Second division.

Compound; of two ingredients.

A. A mixture of diallage and felspar.
- a. With platy felspar.
- b. With fine granular felspar.
- c. With compact felspar.

B. Diallage and actinolite.

C. Diallage and talc, or chlorite.

D. Diallage and serpentine.

Third division.

Compound: of three ingredients.

A. Diallage, felspar, and mica.

B. Diallage, felspar, and quartz.

Memoir on the Mountain of Rock Salt at Cardona, in Spain. By M. P. Louis Cordier.

(Annales des Mines for 1817.)

* *Extract.*

THE description of the rock salt mountain of Cardona, should equally interest natural philosophers and mineralogists. The perfectly insular state of this mountain, its great mass, its peculiar forms, the actual position of the beds of pure muriate of soda, of which it is almost exclusively composed, are, without doubt, well worthy of attention; but the most remarkable circumstance apparently is, that such a mountain, exposed since its formation, to the inclemency of the weather, should have resisted it to this time, and should not have sensibly diminished in size from the earliest records. Its existence does not agree, it must be confessed, with the common hypotheses, which suppose that high mountains, and in general all the inequalities on the surface of the earth, are subject to a rapid decrease. This mountain is moreover as much celebrated in a picturesque point of view, as it is little known in a scientific respect. It has always been regarded as one of the most singular curiosities of Spain; nevertheless it has been visited but by few mineralogists, and the best notice that we as yet have of it, is that

* This memoir, read to the Societé Philomathique of Paris, March 2, 1816, is printed in the Journal de Physique, vol. 82, p. 343 to 358.

given by Bowles (in 1775), in his work, entitled, "Introduccïon a la Historia Natural y a la Geografia fisica de Espanna."

The small town of Cardona, which has given its name to the mountain of salt, is situated in the interior of Catalonia, sixteen leagues from Barcelona, and seven leagues from the central ridge of the Pyrenean chain. It is built upon a platform constituting part of the heights that border the right bank of the Cardonero, a small river flowing towards the south. From the barometrical observations of many days, calculated with corresponding notes made at Barcelona, M. Cordier found that the soil of the town of Cardona, at the foot of the walls of the castle, on the eastern side, was elevated 411 metres [1,348 feet] above the Mediterranean. He moreover observed that the same spot was 138 metres [452 feet] above the mean height of the waters of the river in that part of the valley.

The elevation and commanding position of the castle of Cardona, make it a favourable station for observation. A person placed on its walls, sees without obstruction over an immense extent of low mountains, all of secondary formation. These rocks rise, on the north, gradually towards the highest crests of the Pyrenees; on the east, they disappear at a distance under the platforms of San Miguel del Fay, which are entirely calcareous; on the south, they stretch more than ten leagues, and serve as a base to the singular system of sandstone and puddingstone rocks, that compose the insulated mass and grotesques passes of Mont Serat; lastly, they extend towards the west, and compose a part of the elevated surface of Aragon.

The salt mountain appears as an outwork in the midst of this vast extent of country; the observer sees it in some measure beneath his feet on the south east; he recognizes it by its insular character, its sharp forms, by the hollows (effondremens) partly surrounding its base, and above all by its red and white colours, the vivacity of which contrasts with the grey and sallow tints of the secondary rocks. These rocks form an inclosure round the mountain in the shape of a horse-shoe, which opens towards the east into the valley of the Cardonero, and whose short axis is nearly from east to west. The town and castle of Cardona are situated at the extremity of the northern branch of this horse-shoe.

The inclosure is about three kilometres [9837 feet] long, and one kilometre [3279 feet] broad. Its circumference almost every where affords rapid slopes, or abrupt escarpments; its edge presents slight inequalities, the elevation of which differs little from that of the town of Cardona.

The mountain of salt occupies about two-thirds of the area of the circus, beginning at the back part; its height, above the Cardonero, scarcely exceeds 100 metres [328 feet]; so that its mass is hardly higher, or broader than Montmartre, near Paris, although it is about one third longer.

Its general form is that of an irregular mass, elongated in the shape of an ass's back, presenting escarpments more or less abrupt in many places. The superior portion is bristled by numerous projections, sharp points, and crests: many of the slopes are here and there encumbered by earthy matters; others, more uncovered, present inequalities a little less defined than those of the summit. In order finally to make this peculiar configuration better understood, it may be added, that it bears a great resemblance to plans in relief of the high Alps.

The bottom of the semi-circular basin, separating the mountain from the sides of the circus, presents numerous inequalities and hollows (effondremens) of greater or less extent; here and there are seen the ruins of secondary rocks, heaps of soft clay, and sharp ridges of gypsum or rock salt.

During rains, the greater part of the waters of the circus form two small torrents on each side the mountain, which unite on the eastern side, and flow into the Cardonero. The remainder of the rain water flows in an opposite direction and loses itself in a vast hollow (effondrement) situated on the north west in the farthest part of the inclosure.

The almost total absence of vegetation on the mountain facilitates the study of its composition and structure; it will not be useless to add, that M. Cordier visited this place in the best season for observation, that is, at the beginning of winter.

These first data being laid down, we shall enumerate the rocks composing the mountain: they may be arranged in six principal divisions, viz.

1. Rock of perfectly pure muriate of soda, very large grained, semi-transparent, and colourless. The grains are quite clear; some are found so large, that, by mechanical division, cubes of two decimetres [nearly 8 in.] may be formed.

2. Pure muriate of soda, the mass small grained, more or less translucent. Its principal colours are greyish white, pearl grey, reddish white, flesh red, wine lee red, and brownish red.

3. Impure muriate of soda, mass granular, which would enter into the preceding division, were it not mixed more or less abundantly with either grey or blueish clay, or with very small white or reddish crystals of common gypsum. This last mixture gives a porphyritic structure to the mass.

4. Grey or bluish clay; it is sometimes pure and slightly schistose, at others porphyritic from the mixture of a great quantity of small common gypsum crystals, that are sometimes grey and opaque, at others colourless and transparent.

5. Common gypsum, the mass small grained; it is opaque; its white colour often approaches grey or yellowish: clay disseminated in small quantities is occasionally found in it. Some small rare grains of grey lamellar carbonate of lime also occur.

6. Common gypsum, mixed with anhydrous gypsum; the mass is granular passing into compact; this kind in other respects resembles the preceding.

These different materials occur in very unequal proportions. The small grained pure muriate of soda (No. 2) may be estimated alone to form seven-tenths of the mountain. The impure muriate of soda and clay each constitute about two-tenths. The gypsum and perfectly pure rock salt (No. 1) amount scarcely to a tenth.

This mode of composition is without doubt worth attention, but the stratification is still more so; thus all the materials of the mountain are disposed in parallel and vertical beds, having a direction from E.N.E. to W.S.W. that is, according to that of the crest stretching through the middle of the circus.

The mean thickness of each of these beds, thus placed on their edges, may vary from one to six decimetres [about 4 to 24 inches]; some are found of not more than a centimetre [$\frac{1}{25}$ inch], whilst others attain a thickness of seven or eight metres, [about 23 to $26\frac{1}{2}$ feet]. Many saline strata of the same sort are often in contact; they then can only be distinguished by their grain and the contrast of their colours.

This order of stratification is variously modified; sometimes the thickness of a bed varies considerably in different parts of its course, thus presenting exact parallelism; sometimes the planes are shifted in an opposite direction, producing various inflexions in the lines of direction and dip. These irregularities do not alter the general order.

No order has been observed in the intercalation of the beds of different kinds. They appear to alternate with each other in no settled manner. All that may be said is, that the clay is most abundant on the northern side, and that its opposite side contains scarcely any thing but muriate of soda. The gypsum beds are not mixed with the rock salt; they occur interposed between the last clay beds on the north side.

Some of the steepest slopes of the mountain are cut by fissures, so wide that they may be entered to the depth of some metres. In these rugged places saline concretions are found, sometimes tubercular, at others in mamillated plates; they produce a fine effect from the contrast their brilliant white colour presents to the tints of the coloured bases over which they are scattered.

The perfectly pure and transparent rock salt beds are almost all united on the E.S.E. side of the mountain; they there form two appendices of little elevation, which are only remarkable because they constitute what is properly termed the salt mines of Cardona, and are the seat of the works carried on by the Spanish government.

These works are the more important as they require but little expense. We shall present an idea of them in a few words.

The workings are carried on in open day, by horizontal cuttings in the shape of steps. Every step is one metre [about 3 ft. 3 in.] high, with a similar width. Their length is great enough to allow of ten miners working in the same line. Eight cuttings of this kind are formed one above the other. The rock salt is first blasted by gunpowder, and afterwards finished with the pick axe. The moderately sized parts only are carried away, and are ground in a neighbouring house. The salt, after having been washed, is, without any other preparation, sent to the government magazine. It is stated that salt to the value nearly of one million of francs is annually sold. Without entering into more details, with regard to the advantages derived from the workings of rock salt, we shall again take up the description of the mountain. It has as yet been considered as insulated: it must now be considered with regard to its relations with the secondary rocks that surround it.

The stratification of these rocks is not less easy to observe, less evident, and by contrast less remarkable than that of the salt mountain. All parts of the secondary strata rise towards the centre of the inclosure, that is, those on the north under an angle of near 50°, and those on the east and south under an angle from 20° to 30°; so that supposing them to be prolonged, they would cover the vertical edges of the saline and gypsum beds.

It is necessary to walk along the foot of the escarpments in the circus, in order to observe the immediate superposition. The base of the salt mountain is seen in numerous situations to dip and disappear under the strata of the secondary formation; the heaps of debris of these same strata elsewhere shew that they have fallen and crumbled in consequence of the successive destruction of their original points of support. The superposition is the less equivocal, as whenever it may be verified, the strata of the two rocks are seen tending to cut each other at angles more or less approaching a right angle.

The composition of the secondary rocks presents a contrast equally striking. The following may be distinguished:

1. Micaceous sandstones of a grey colour, composed in a great measure of large fragments of quartz and granitic slaty rocks; they are very hard, and well cemented.

2. Red fine grained micaceous sandstone, possessing a very compact texture.

3. Red, green, and grey argillaceous schist, commonly covered with small scales of grey or white mica, placed in the direction of the laminæ.

4. Hardened schistose clays, sometimes altogether soft; they are greyish or greenish white, or even reddish brown.

5. Compact limestone, of a scaly fracture, of a dark grey colour, sometimes a little greenish; it is often mixed with portions of green schist, and suddenly with some particles of mica. It has no bituminous odour; M. Cordier was unable to discover any traces of marine bodies in it, not only in the environs of Cardona, but also in the other places in Catalonia where it occurs. It is nevertheless probable that it contains them, though they are extremely rare.

6. Argillaceous limestone, of a grey or greenish colour, often abounding in bunches of mica, without bituminous smell, shelley, and affording but rarely very slight remains of carbonized vegetables.

These different rocks alternate indifferently with each other, but in such a manner that the sandstones predominate in the inferior part of the system, and the limestone in the upper.

Thus, as has been said, these rocks not only constitute the environs of Cardona, but also a great part of the surface of Catalonia. They occur every where with the same characters, but affect variable dips, and opposite directions; these dips rarely exceed 30° or 40°; the most highly inclined, that M. Cordier observed, are seen at Suria, a village situated on the left bank of the Cardonero, two myriameters [about 12½ miles] below Cardona; the limestone beds rise at 70° towards the N. N. E.; they possess the peculiarity of containing a bed of poor coal, one metre [about 3 ft. 3 in.] thick, parallel with them.

From the characters presented by the system of rocks covering the base of the salt mountain of Cardona, M. Cordier considers it as belonging to the most ancient formation of the secondary rocks.

Considering moreover that the strata composing this system are not conformable to those of the salt mountain, that on the contrary they cut each other at nearly right angles, or as separate formations, that the superposition is evidently non-conformable (transgressive), he concludes that the salt and gypsum rocks of Cardona belong not only to an anterior, but also to an absolutely distinct formation, and that it ought to be regarded as one of the intermediate (transition) series.

In order to strengthen this important conclusion, M. Cordier quotes the results of the observations he made in 1804 and 1809, on the position of the gypsum rocks of Mont Cenis and the Little St. Bernard; results tending to prove that these rocks, sometimes composed of common gypsum, sometimes of anhydrous, at others épigène, or mixed with nitrate of soda, form true beds, often very thick, which are incontestably subordinate to the transition series, which forms a large portion of that part of the High Alps.*

M. Cordier afterwards treats of the imperceptible decrease of the salt mountain of Cardona; he examines the confirmed

* In a subsequent paper by M. Brochant de Villiers, will be found a detailed account of the gypsum in the Alps, from which there are very strong reasons for supposing that many, if not all of them, belong to the saliferous, or new red sandstone series, and are therefore secondary, though the author does not consider them to be of that epoch. (Trans.)

prejudices of the country on the subject. He estimates, from some experiments, that the rain water flowing down the sides of the mountain, ought rarely to acquire a saltness greater than 4°. He finds that the specific gravity of the rock salt of Cardona, in cubic and clear pieces, determined with the essential oil of turpentine, is 22.1967 (that of distilled water being 10). Lastly, supposing that there annually falls at Cardona, eight decimetres of rain, and admitting that each decimetre acquires a saltness of 4°, he determines, by a simple calculation, that the eight centimetres ought annually to carry away from the upper parts of the saline mass a thickness of 15 millimetres, 26 hundredths of rock salt; whence it follows that the mountain only diminishes 152 centimetres 6 tenths (5 feet, 0 inches, 8 lines English) in a century. M. Cordier only offers this as a fair sketch, from which it may easily be understood why the decrease of the mountain of Cardona has always appeared imperceptible.

We shall finish with the following summary of the principal geological results of M. Cordier's memoir.

1. The saline and gypsum rocks of Cardona occur in vertical beds.

2. This system is covered by the most ancient secondary beds, whose superposition is non-conformable (transgressive.)

3. From the nature of this superposition, the gypsum and salt beds are, without doubt, of an era, not only anterior to, but altogether distinct from, that of the secondary beds.

4. Pure and occasionally saliferous gypsum occurs in the High Alps, and incontestably forms part of the transition series.*

* The saliferous gypsum of Bex, so long considered as a decided example of the occurrence of such rocks in the transition series, has been shewn by Professor Buckland to belong to the saliferous, or new red sandstone formation; it would be much more in accordance with what we know of the general geological position of rock salt, to consider that

5. These saliferous gypsum rocks possess marked analogies with the system of Cardona.

6. From the preceding data, this system ought itself to be considered as forming part of the transition series.

7. Hence it follows, in the last place, that another formation of rock salt and gypsum, must be admitted in geological arrangements.*

of the mountain of Cardona as constituting part of the saliferous, or new red sandstone formation, which surrounds it. It is possible that the vertical position of the salt beds may be occasioned by some contortion in that formation, as the beds on either side rise towards the ridge of the mountain. The only fact against this hypothesis is the non-conformable position of the salt beds and new red sandstone at those places in the valley where they may be observed in contact. (Trans.)

* An account of the salt mines at Cardona has been given by Dr. Traill, in the 3d volume of the Geological Transactions, p. 404 to 412. (Trans.)

Observations on the formations of Ancient Gypsum occurring in the Alps, particularly on those considered as primitive; preceded by new facts relative to the transition rocks of that chain. By M. Brochant de Villiers, Engineer in Chief of the Royal Mining Corps.*

(Annales des Mines for 1817).

Read at the Royal Academy of Sciences, March 11, 1816.

[THE author commences this memoir by observing, that geology had made rapid advances since Werner had first called the attention of geologists to the different rock formations; that nevertheless the formations originally traced by Werner, had been subsequently modified by himself, his pupils, and others; that the study of shells, and of fossil organized bodies in general, had tended greatly to advance our knowledge of secondary rocks; that less attention had of late been paid to the primitive rocks; but that the transition series had been the object of numerous researches. He mentions Messrs. Von Buch, Haussman, Brongniart, Omalius, Raumer, and Bonnard, as the principal naturalists who have thrown light on this class.

M. Brochant de Villiers states at the same time, that before their observations were known, he had also applied himself to this branch of enquiry, and that in a memoir

* The author appears to use the words "Ancient gypsum," in contradistinction to the Tertiary gypsum. (Trans).

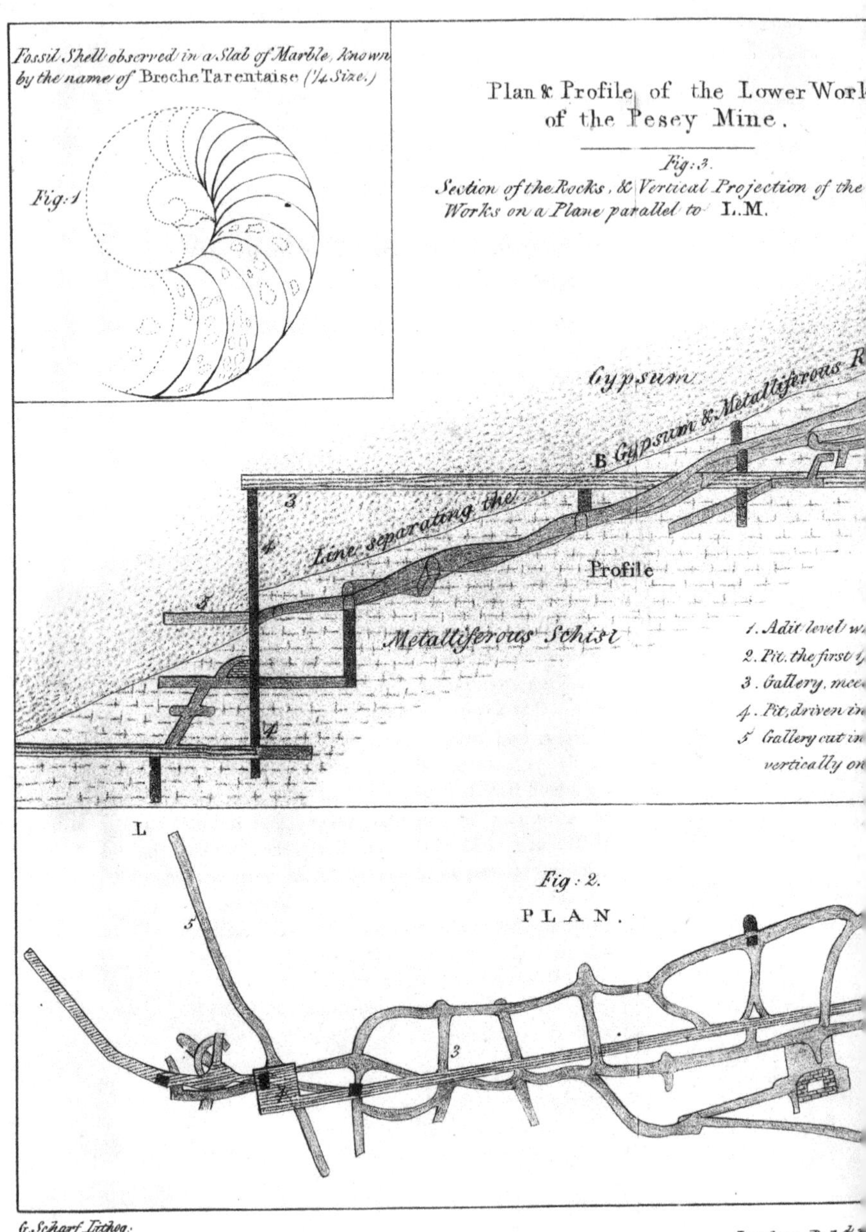
Fossil Shell observed in a Slab of Marble, known by the name of Breche Tarentaise (¼ Size.)
Fig: 1
Plan & Profile of the Lower Work
of the Pesey Mine.
Fig: 3.
Section of the Rocks, & Vertical Projection of the
Works on a Plane parallel to L.M.
Gypsum
B Gypsum & Metalliferous R
Line separating the
Profile
Metalliferous Schist
1. Adit level w
2. Pit, the first
3. Gallery, mee
4. Pit, driven in
5 Gallery cut in
vertically on
L
Fig: 2.
PLAN.
G. Scharf Lithog.
London. Pub.d

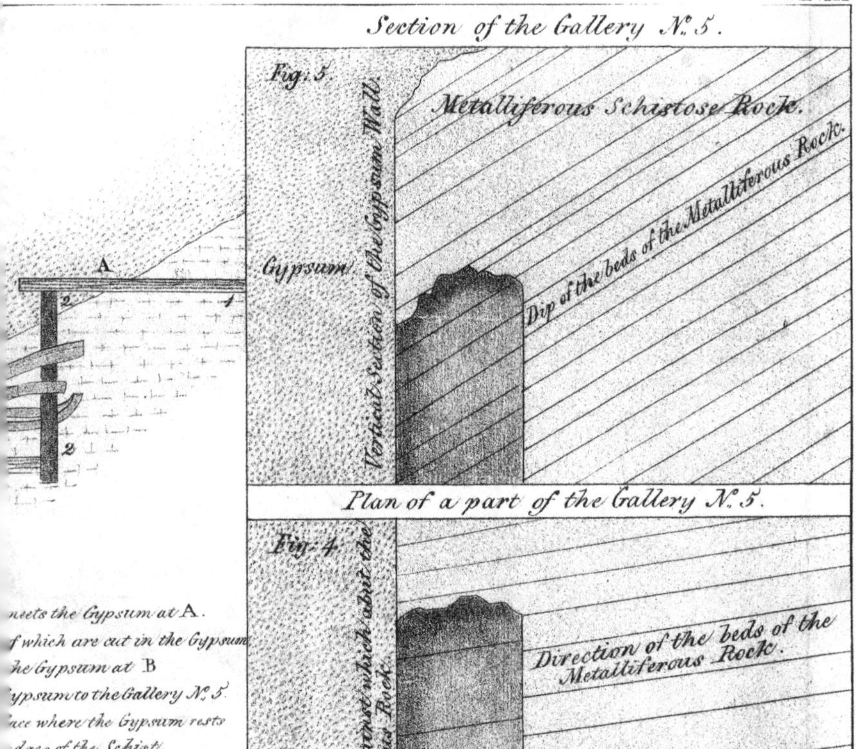

Section of the Gallery N.° 5.

Fig: 5

Metalliferous Schistose Rock.

Dip of the beds of the Metalliferous Rock.

Gypsum

Vertical Section of the Gypsum Wall.

A

Plan of a part of the Gallery N.° 5.

Fig: 4

...neets the Gypsum at A.
...f which are cut in the Gypsum
...he Gypsum at B
...ypsum to the Gallery N.° 5.
...ace where the Gypsum rests
...dges of the Schist.

Direction of the beds of the Metalliferous Rock.

Horizontal Section of the Wall of Gypsum against which abut the edges of the beds of the Metalliferous Rock.

Gypsum

M

Metalliferous Schistose Rock.

Printed by C. Hullmandel.

Phillips, 1824.

MAP of the VAL CANA

with two Sections: shewing the connexion of the Rocks of the

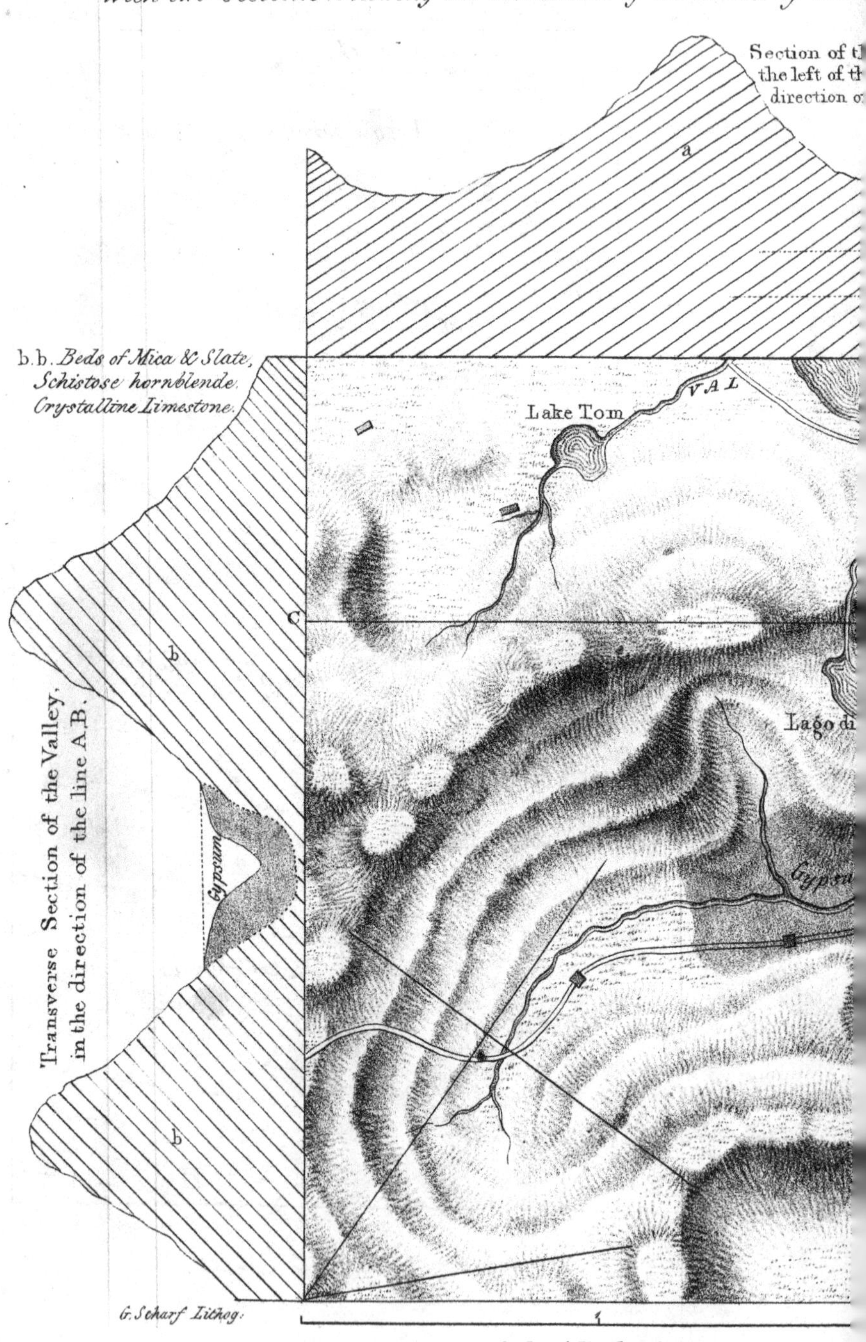

IA near St GOTHARD. PL : IV.

ntains forming the Valley, with the Gypsum at the bottom of it.

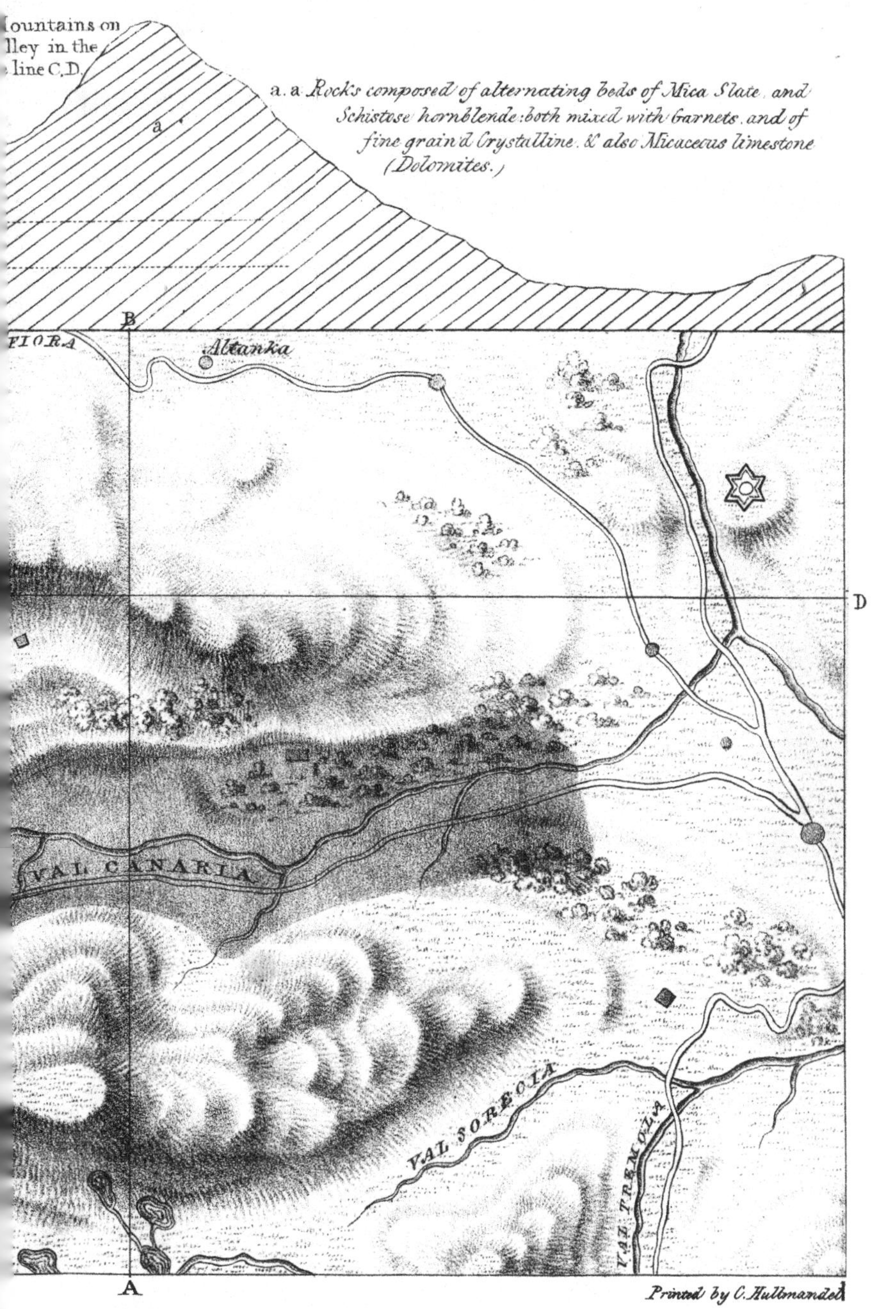

Printed by C. Hullmandel

ndon: Pub.^d by W. Phillips, 1824.

read before the Institute in 1807,* he had shewn that many transition rocks existed in the Alps, and more particularly in that part of Savoy known by the name of Tarentaise, affording characters different from those that had at that time been observed in Germany, and containing rocks that had until then been arranged exclusively in the primitive series.

He observes that every subsequent examination had confirmed him in his first ideas, which had also been approved of by Messrs. Von Buch and Omalius d'Halloy, who had visited the country.

Notwithstanding his conclusions had been adopted by many celebrated geologists, there were others who still doubted them, in consequence of the total absence of organized remains in the crystalline limestones, which form a considerable portion of these transition rocks.

The author then states that all the rocks of the Tarentaise were subordinate to two divisions, the limestone formation, and the puddingstone containing anthracite, or anthracite formation ; in the latter, vegetable impressions were found, but no organized remains had been discovered in the former ; his principal proof of its belonging to the transition series, was its alternation with the anthracite formation.† He was most anxious to obtain the last decisive proof (the occurrence of organized bodies) of its belonging to that class ; this however he sought for in vain for a considerable time, when most unexpectedly, he discovered at Paris, with the assistance of M. Leman, a large shell of the nautilite or ammonite kind (plate 2, fig. 1) in a table of the Breche Tarentaise marble, a rock found in the midst of the district, containing his transition rocks.

He then proceeds to describe the gypsum occurring in the Alps].

* This memoir is inserted in the Journal des Mines, vol. 23, p. 321.

† As the lias, alpine limestone, and new red sandstone formations, are so much altered in the Alps, may this anthracite formation be the representative of the coal measures? an attentive examination of the vegetable impressions would probably decide this question. (Trans.)

On the ancient Gypsum of the Alps.

The same part of the Alps in which I have observed these transition rocks, also contains gypsum, often in very large masses. I have in the description of those rocks, taken care to mention them; but having at that time, notwithstanding many years of observation, some doubts on the true position of these gypsums, and on their relation to the transition rocks, I abstained from pronouncing an opinion on them, reserving the clearing up of their geological characters for another memoir, when I should have been able to make some important verifications.

The following are the principal points on which my doubts were founded. The gypsum rocks are generally extremely crumbled, the natural consequence of the great facility with which this substance decomposes; there necessarily results a great difficulty in observing their position relatively to the other surrounding rocks.

Their situation in the Tarentaise, on the sides of mountains, or on their first escarpments, often even at the bottom of high valleys, generally in masses of little extent, and always superficial, inclined me to presume that the gypsum was a formation posterior to all the rocks of the Tarentaise, that is, even to the transition series.

This conjecture, moreover, appeared to me to agree with the ideas, without doubt rather vague, that a number of philosophers had entertained on gypsum in general, according to which it would always be deposited in basins.

I nevertheless abstained from decidedly adopting this conclusion; besides its not appearing to me sufficiently proved, I had good reasons for remaining in doubt. In the first place, gypsum had been met with in subterranean works of the mine of Pesey; I presumed that it was covered by the metalliferous rock of a later formation, but of this I was not completely assured. Many persons held a contrary opinion.

Gypsum also existed in many other places in the Alps; mineralogists mentioned many as belonging to the transition series; and they all agreed in admitting the gypsum of St. Gothard to be primitive; celebrated geologists had visited it, and had considered it as such. M. Daubuisson had observed the gypsum of the valley of Aosta, which he considered also as primitive; so that in all mineralogical and geological works, a gypsum formation was seen placed in the class of primitive rocks.

It was not altogether absurd to admit among the gypsum of the Alps, primitive and transition gypsum, and at the same time another gypsum, very ancient, but of later formation; the two first divisions have indeed been generally admitted. But I had seen specimens of all these gypsums, and, on comparing them together, I found so much resemblance, so many relations, and, if I may so express myself, such a striking family air, that I could not determine to allow them different origins; on the other hand, it was impossible for me to agree with the idea of primitive formation, which was attributed to many of them, from the geological characters afforded me by that of Tarentaise. I had briefly exposed all these doubts in the memoir on the transition series; I have since further developed them in a note read before the Société Philomathique, which has not been printed: I shall now occupy myself with their solution. It will be seen that my new researches have led me to conjectures differing a little from my first, but at the same time I conceived myself obliged not to admit the primitive gypsums against the opinion now adopted by all geologists, even those most celebrated.

I had already, in 1809, visited the gypsum of the valley of Cogne, announced as primitive by M. Daubuisson; I had the advantage of having him as a guide in this excursion; but I was thwarted by the great abundance of snow that had fallen in the preceding winter, the melting of which had been much retarded. The gypsum was covered by it, and I could only observe the surrounding rocks.

In 1813, the order I received to conduct the students of the mines in a geological tour in the Alps, appeared a favourable opportunity to me; being entirely master of my route, I directed it principally to many points, where I knew gypsum existed, and more particularly to those that had been noticed as primitive.

From Moutiers in Savoy, where the practical mining school then was, I passed the central chain of the Alps by the Col de la Seigne, and the Allée Blanche. I descended into the valley of Aosta, from whence I visited that of Cogne; afterwards crossing the mountains which form as it were the advanced posts of the Alps on the side of Piedmont, I reached St. Gothard, from whence, repassing the chain, I followed the Valais to Bex. I observed gypsum during this tour.

1st. In the Allée Blanche.

2ndly. In the valley of Cogne; that of M. Daubuisson.

3rdly. In the Val Canaria, at the foot of St. Gothard; that of Messrs. Freisleben and Von Buch.

4thly. Near Brigg, in the Valais.

5thly. At St. Léonard, near Sion.

6thy. At Sarran, near Martigny.

7thy. Lastly, at Bex.

It is from comparing the mineralogical and geological characters of these gypsums, and at the same time of those that I had previously observed in Savoy, and of many others of which I could collect descriptions, that I believe myself able to establish that all belong to one period of formation, or at least to two epochs of that period, which is that of the transition series.*

I shall consider more at length those which have been regarded as primitive, so as to expose my motives for not adhering to that opinion.

A. *General characters of the Gypsums of the Alps.*

The greater part of these gypsums are apparently in interior texture rather compact than crystalline; in general,

* It will hereafter be seen that there is some reason for doubting this conclusion, and believing that their true epoch is rather that of the secondary series. (Trans.)

at least, they only present some brilliant points, even with a magnifying glass; but if they are coarsely powdered, the brilliant points are multiplied to the eye, and the powder observed with a microscope shews, even in its smallest particles, only crystals of sulphate of lime in rhomboidal tables.

Besides the small crystals, which appear to form the mass, larger crystals of sulphate of lime are seen imbedded in it. There are nevertheless some gypsums which appeared to me decidedly compact, but their colour was grey; they are however associated with crystalline gypsum.

Some varieties are a little laminated, but the greater number do not shew this structure in a decided manner. Their masses break indifferently in every direction.

These gypsums are generally of a snow white colour. When considerable masses are met with, free from mixture, they are advantageously employed in sculpture, as is the gypseous alabaster of Tuscany. (That of St. Leonard has, it is said, been used to make statues for the churches at Fribourg). I have found the grey varieties only in Savoy, in the Val d'Arbonne near St. Maurice, and the yellow near Brides; this last colour appears to be owing to subsequent alterations; all the surrounding gypsum is white.

Many different mixtures are seen in it.

1st. Carbonate of lime, which is compact, sometimes of a dull grey colour, at others and most frequently of a dark blackish grey. It exists under the remarkable form of nodules, most commonly angular, and sometimes rounded. The greatest number of mineralogists, who have observed these mixtures, have considered them as conglomerates or arenaceous rocks, which gives the nodules an existence anterior to the paste of gypsum containing them.

I had myself embraced this opinion, and do not pretend to regret it yet; but the workings at Sarran have led me to doubt the pre-existence of some of these nodules. Small veins of compact limestone are found in this gypsum quite parallel with the beds, and having all the characters of contemporaneous origin. I should be led to suspect, especially from some characters of position, that the well recognized nodules were of an origin little anterior to that of the gypsum.

These decided calcareous nodules occur abundantly in the gypsum of Pesey, also in that of Brides in Savoy. Those of St. Leonard and Bex afford it also. They are perfectly decided in the two first places.

2d. Mica or rather talc. It is only found in two of the above mentioned positions; at the Val Canaria and Brigg. At the Val Canaria, mica is disseminated in an uniform manner in the rock, sometimes in little insulated plates of a greenish yellow, at others, a little fibrous in small groups. Even in the midst of the gypsum some yellowish masses are observable, that are principally composed of mica; they also contain gypsum, but it is as it were dissolved in the midst of the mica, the colour of which it has taken, and to which it serves, it may be said, as a paste. These masses afford shining surfaces, when broken, in which the mica is fibrous.

I have here used the name mica, because the gypsum of that locality has been generally noticed as micaceous; but I should observe, that it possesses many of the characters of talc, that it is one of those numerous passages, recognized between these two substances, which have made their future union into one mineralogical species probable, passages of which the rocks of the Alps have afforded me many examples.

At Brigg, the talc is more decidedly shewn. It is not disseminated throughout the mass of gypsum as at the Val Canaria; it forms plates or rather a sort of varnish on the surface of the laminæ of the rock which is a little schistose.

Its colour is a silvery greenish white; it is slightly fibrous. This presence of mica and talc in these two gypsums, joined with the identity of colour of the gypsum itself, is certainly not sufficient to assign them the same geological origin; nevertheless, it cannot be denied that there exists between them a sort of resemblance, which is the more remarkable and extraordinary, as the last of these gypsums (that of Brigg) is referred to the transition series by many mineralogists, who regard the first (that of Val Canaria) as primitive.

3d. Steatite. This mixture is principally seen in the gypsum of Cogne and in that of Sarran near Martigny.

In the first, steatite forms small masses, most frequently flat, and even small thin plates of a leek green colour, often blackish, very translucent at the edges, and sometimes of a fibrous texture; these plates are disposed parallel to each other, and to the plane of the gypsum bed, but they do not form continuous planes, and do not give a laminated texture to the rock; they are too much scattered, and often of rare occurrence.

In the gypsum of Sarran, the steatite is in smaller plates, of not so dark a green, but duller, and of a texture often fibrous, approaching talc. They are disposed in the same manner, generally closer together; but there are also large portions of gypsum which do not contain any at all. There are among these plates some that bear a great resemblance to the superficial thin plates of talc, ill defined at the edges, mentioned above in the gypsum of Brigg.

The gypsum of St. Leonard also affords some small plates analogous to fibrous steatite, but ill determined.

Lastly, I obtained a specimen of gypsum found at St. Gervais, in Faucigny, near the hot springs, that contains a great number of these plates of steatite, which is fibrous, and often shining, without however possessing the demi-metallic lustre of talc. I have not myself observed this gypsum in place, but I have many times visited the country, and different observations make me regard it as altogether analogous to the gypsum of the Tarentaise, which is in the neighbourhood.

I shall here remark the affinity which the mixture of steatite shows between the gypsum of Cogne, one of those regarded as primitive, and many others (those of St. Leonard and Martigny) considered as transition, not only according to my own ideas, but according to those of other mineralogists who admit the primitive gypsums of the Alps.

4th. Anhydrous sulphate of lime. This substance has been equally found in many gypsums of the Alps, and in gypsums of Germany, which belong to very ancient rocks, but which are recognised as secondary. It exists in the Alps in the midst of gypsum masses and also in the rocks near them. Defined crystals and crystalline masses with large laminæ more particularly occupy the clefts in the neighbouring rocks; but in the midst of the gypsum, the anhydrous sulphate of lime more commonly forms bunches, and more or less considerable masses of a confused laminated structure, with a crystalline fracture, sometimes in concentric layers, of a whitish grey colour approaching violet.

I have thus observed it at the glacier of Gebrulaz, in the gypsum of Pesey, in the Tarentaise, at Allevard, &c.; I do not however pretend to affirm that this substance never forms beds in the gypsum; it certainly appears to occur thus at Bex.

I am not as yet acquainted with any description of the position of the anhydrous sulphate of lime of Vulpino. If, as I am led to presume, it is analogous, the masses that it forms in the gypsum must be very considerable.

5th. Muriate of Soda and saline springs. The presence of muriate of soda is recognised in many of the gypsums of the Alps, that I have observed. The gypsum contains rock salt in the valley of the Arbonne, near St. Maurice in the Tarentaise; this salt is often visible in powder on the surface; it also has been attempted, and not without success, to obtain muriate of soda from it by solution.

The saline spring of Moutiers rises out of limestone, but at the foot of a considerable mass of gypsum; other saline springs of the Alps are in the gypsum itself, which often contains a little rock salt. The gypsum of Bex appears to be decidedly inclosed in transition rocks;* but on the other hand, many saline springs and deposits of rock salt worked in Germany occur in masses of gypsum mixed with clay, or, as has with truth been said, in masses of gypseous muriatife-

* It will be hereafter shewn that they are secondary. (Trans.)

rous clay; and these latter gypsums are known to be generally considered as of decidedly secondary formation.

These resemblances are sufficiently remarkable; I do not pretend to conclude from it an identity of geological formation between these ancient secondary gypsums of Germany and the gypsums of the Alps, even those in insulated superficial masses; but they may one day serve to establish relations between the different terms of the series of rocks, the formation of which has followed that of the primitive. I shall hereafter return to the subject.

6th. Sulphur also exists in the gypsum of the Alps; it forms nests of rare occurrence, and of small size, in the mass of the gypsum. It has been found in that of Bex, at Pesey, at Gébrulaz, &c.: there occurs at Allevard a lamellar anhydrous sulphate of lime, which is penetrated by sulphur, giving it a yellowish tint.

I might also notice anthracite, some traces of which are found in the gypsum of Brides near Moutiers; but from this example a very positive consequence cannot be drawn as to the identity of the position of the gypsum and anthracite, this substance only occurring in very thin small veins, between the laminæ of the gypsum, in portions near the surface: which may lead to the presumption that it has been afterwards deposited by infiltrations from the water of the upper rocks, which contains large masses of anthracite. It will hereafter be seen that I have obtained more probable data of the existence of the gypsum of the Alps and anthracite in the same formation.

B. *Geological positions of the Gypsums of the Alps.*

1. In the Tarentaise (Savoy).

I have always observed the gypsum of the Tarentaise on the surface, and as I have already said, in such a state of decomposition that it was impossible to judge, in a rigorous manner, of its geological relation to the other rocks. It occurs on the sides of mountains, or sometimes forms white slopes resembling snow, (valley of St. Bon.)*

It is also seen on the summits, at least on the crest of the first escarpments. But what is very remarkable in a country where the beds are always highly inclined, it is not seen to descend lower, excepting the fallen masses which are evidently out of place; they do not therefore form part of the mass of the rock. I nevertheless conceived that I had several times discovered beds of gypsum near Moutiers, sometimes in the midst of transition limestone, at others in the midst of the anthracite formation; yet notwithstanding my researches I have never been able to acquire a positive proof of it; the rocks which give rise to this conjecture being too much fallen over, and too much covered with vegetable earth: I should now consider this conjecture more probable from the observations made in Piedmont and the Valais, during my tour in 1813, of which I shall hereafter give an account.

These summits, crowned by gypsum, never attain a greater elevation than from 2,000 to 2,400 metres † [about 6562 to 7874 feet]. They are not actually the summits, but the first escarpments of the mountains, serving as a base to the high valleys, the region of pastures. In the bottom of these high valleys masses of gypsum are also sometimes found, but, are always superficial, ‡ and not in sufficient quantity to give a complete idea of their formation in basins; an idea that I do not pretend to adopt, but which almost inevitably presents itself when these gypsums are met with, and are seen occupying the bottoms of ancient high valleys. I shall have occasion to cite a more striking example. ‖

The points where gypsum is found in the Tarentaise are comprised in the valleys where the anthracite formation abounds, often with vegetable impressions; § I at first paid

* Corresponding appearances have been remarked with regard to the gypsum of Salzburg.

† St. Bon, Champagny, Croix de Fessons.

‡ Near the Lake de Tines, Gebrulaz, and Pesey.

‖ In the Val Canaria.

§ Valley of Bosel, Brides, Champagny, Gebrulaz, St. Bon, Moutiers, Valley of Pesey, and the Val d'Arbonne opposite.

little attention to this circumstance, which now becomes very important from the existence of gypsum on the Allée Blanche, of a deposit of gypsum in the Valais (that of St. Leonhard) on an anthracite formation, and of another (that of Brigg) in a formation that is analogous to it.

Lastly, as I have already announced, gypsum is found even in the works of the mine of Pesey, and this is the place to discuss the true characters of this position, because it is very important.

There is no question here of the small portions of anhydrous sulphate of lime, that are sometimes met with in the mass even of the ore or metalliferous rock, and which have been collected in cabinets. They are easily admitted to be owing to infiltrations from the upper parts; I mean to speak of the considerable masses of gypsum and anhydrous sulphate of lime, the debris of which cover the refuse of the mines (haldes).

These gypsum masses do not exist in the works carried on for the ore, they do not contain a trace of it; they are only met with in the adit level, and solely in the portion of this gallery above the useful works. The gypsum is then evidently above the metalliferous rock. This position has been fully admitted; but it may equally happen in two very different cases: either the gypsum there forms a distinct bed belonging to the same system as the rock, or else it has been deposited upon it at a period subsequent to its consolidation, and constitutes part of a different formation.

The surface cannot afford any information in this respect, being entirely covered by tuffa, pasturage, and wood, the gypsum not shewing itself. It is met with only at the hamlet of Beaupraz, at 1200 metres [3937 ft.] distance, ascending the valley; it occurs there at the surface, forming an insulated mass of small extent, at the foot of one of the slopes, where it loses itself under the vegetable soil.

The other characters that I have noticed above the Tarentaise gypsum led me already to presume that this gypsum covered the rock, that its formation was more modern, and consequently posterior to the transition limestone, since I have elsewhere shewn that the metalliferous rock of Pesey, which is an argillo-steaschist, was of the same formation as all the limestone of the Tarentaise.

In order more fully to assure myself of it, I searched the points of the adit level, and the excavations that communicate with it, which led from the rock to the gypsum, in order to observe their superposition; but the wood work and the decomposition of the two rocks prevented me from obtaining any decisive result. I remained therefore in the same uncertainty with regard to my first presumption, which nevertheless always appeared to me the most probable opinion.

Lastly, while following the workings the miners met with gypsum, and in order to re-discover the metalliferous rock they were obliged to abandon the principal gallery and sink a shaft. I observed the spot where the metalliferous rock was in contact with the gypsum; the latter was not distinctly stratified, and the laminæ of the metalliferous rock were cut very obliquely; so that their mode of association was still doubtful.

To clear this up, I requested M. Schrieber, at that time director of the practical school and the mine, to drive a transverse gallery in the metalliferous rock, precisely at its junction with the gypsum.

This gallery (which, having geological research for its principal object, was named the geologist's gallery), was executed, and driven the distance of twenty metres [about 65 feet 6 inches]; it furnished me with an opportunity of judging decidedly on the position of the gypsum with regard to the metalliferous rock.

It was found that the gypsum was only vertically placed against the edges of the beds of the metalliferous rock, which were cut vertically. One of the sides of the gallery was a perpendicular wall of very even gypsum, the other presented the edges of the beds of the rock cut a little obliquely to their planes; it then appeared to me evident that the gypsum was a later formation than the metalliferous rock, since it not only covered it, but filled if not a cleft that traversed it, at least a cavity, or hollow on its surface. The annexed plans and sections (plate 2, fig. 2, 3, 4, & 5), represent in an exact manner the different works noticed, and the relative position of the beds of the metalliferous rock and the gypsum.*

It would have been desirable, in order to complete this determination of the position of the gypsum, to continue the principal gallery in the midst of its mass, in a direction perpendicular to the vertical wall noticed above, so as to meet with the other side of this cavity; but the expense might have been very considerable, and the poor state of the ore did not authorise it.

Uniting this last fact with the others before noticed, it is seen that all the characters I have been able to collect upon the gypsum of the Tarentaise, *lead to the conclusion that it is above the transition rocks, and of a more modern formation, at least in general;* for, from the other positions I am about to describe, it is not impossible that gypsum may be discovered even in the midst of this transition series.

2. *Gypsum of the Allée blanche.*

It is on the right slope of this valley that the gypsum occurs; it there forms many white pyramidal masses, two of which are very near the torrent: I saw three others at a distance scattered on the slope to the height of 1000 metres [3281 feet] above the valley, that is to say, an absolute height of 2000 to 2400 metres [6562 to 7874 feet]. I only

* Figs. 2 and 3, plate 2, are the plan and section of that part of the lower workings of the mine of Pesey which is here noticed; the line of demarcation between the gypsum and the schistose metalliferous rock is there traced, as far at least as it could be determined by means of the draining galleries 1 and 3, the pits 2 and 4, and others that cut this line, and lastly by means of the gallery No. 5, which is that of the *geologists*, of which I have spoken.

Figs. 4 and 5, are the plan and section of this gallery, upon a scale twenty times larger. They shew the vertical position of the gypsum with regard to the edges of the beds of the schistose metalliferous rock.

examined the pyramids near the torrent; the largest is from 500 to 600 metres [about 1640 to 1968 feet] in breadth, by about 100 or 120 metres [about 328 to 424 feet] in height. This shape is the more remarkable as it rests upon the edges of beds of a well characterized anthracite formation, * and as the smallest trace of gypsum is not observable in the beds of that rock on either side of each of these pyramids, nor the slightest derangement that may make an association of the two formations presumable.

The gypsum appears here, as in the Tarentaise, in the neighbourhood of rocks containing anthracite, and to be of a formation posterior to it: it may perhaps be conjectured that these pyramids are the scattered remains of a more considerable deposit which might once have filled the valley.

3. *Gypsum of St. Léonard.*

In order to go from Leuck to Sion in the Valais, we pass at pleasure either the village of St. Leonard, which is elevated, or by a lower road on the banks of the Rhone. I knew that gypsum was found in this place, and that it was met with equally on either road.

My expectations were not disappointed; yet I did not find a stratification so regular as to shew the relative position in a decisive manner; but I was struck with the presence of anthracite, and its accompanying black argillaceous schist, in the vicinity of the gypsum, and at the same time with calcareous rocks possessing that true or apparent arenaceous structure, so common in the Tarentaise.

The irregularity of this association, which might be referred to a falling down of the rocks, did not permit me to assign all with certainty to the same formation; it nevertheless afforded a striking resemblance to the gypsum of the Tarentaise and that of the Allée Blanche, which exists in anthracite valleys, and is close to a limestone altogether analogous.

* I found among the fallen masses many pieces of anthracite.

My conjectures were strengthened by what I had the day before observed of the Brigg gypsum, of which I shall speak presently, and which presents a similar position with complete evidence, and I knew moreover that many geologists had referred the St. Léonard gypsum to the transition series, so that I did not consider it necessary to stop at it.

Lastly, what was then but conjecture, has since become a certainty; M. Lardi, who had observed the gypsum of St. Léonard on the heights, having found it evidently associated with transition slate.

4. *Gypsum of Bex.*

I shall say but one word on this gypsum deposit, in which subterranean works have long since been carried on in order to search for the saline springs it contains, and which has already been described by many mineralogists.

I had visited it in my first travels in the Alps; but I revisited it with greater interest in company with M. de Charpentier, who at present directs the works.

I there as elsewhere found great confusion in the position, a confusion that has given rise to so much discussion as to the works; what appears to me very probable, and what has since been confirmed by a letter from M. de Charpentier, is that the gypsum there forms beds in an argillaceous limestone, and that the one and the other form part of the transition series.*

The argillaceous limestone contains some beds of a schistose greywacke,† and there is anthracite in the upper parts.

* Professor Buckland has shewn (Annals of Philosophy, new series, vol. 1, p. 455, &c.) that the saliferous gypsum of Bex, is that of the new red or saliferous sandstone of England, and therefore secondary; lias rests upon it, and it is associated with Alpine limestone, analogous to the English magnesian limestone; the observations I have myself made on the spot, lead me to the same conclusions as Professor Buckland. (Translator.)

† These are not true greywacke beds, but beds belonging to the new red sandstone formation.

Nevertheless, many characters make me presume that these transition rocks are a little more modern than those of the Tarentaise.

With regard to the gypsum of Sarran, near Martigny, I could only see the portion that is worked, all the environs being covered by vegetable soil. I know that it is considered as transition gypsum, but I could not obtain the proof of it; it is very certain that it is not anterior.

5. *Gypsum of Brigg.*

I now come to the two examples in which only I have myself seen the gypsum very evidently associated with other rocks.

At Brigg, or rather at about 2000 metres [6562 feet] N.E. from that town, on the left bank of the Rhone, a well characterised bed of gypsum is seen, almost projecting over the bed of the river. Its direction is the same as the valley of the Rhone, nearly E.N.E. and W.S.W. It dips at 45° to the South, exposing its edge to the valley.

This gypsum is covered by a crystalline whitish grey limestone, schistose and mixed with mica. On this limestone is seen another much more coloured, then a spotted blackish schist, that effervesces; and lastly another schist, that also effervesces, but of a much darker colour, containing insulated plates of mica, and altogether resembling the schists that accompany the anthracite; the whole of the thickness of some metres.

It was then very certain that the gypsum formed the integrant part of a formation, and it might already be conjectured to be a transition formation, from the nature of the last mentioned rocks; this presumption moreover became certain, or at least extremely probable by the other characters of the ground in this part of the valley of the Rhone which belong to this formation; thus this gypsum has been admitted as transition by many mineralogists who have observed it.

6. *Gypsum of Cogne.*

The position of this gypsum, noticed as primitive by M. Daubuisson, who discovered it in 1807, has been described by him in a notice inserted in the Journal des Mines, No. 128, p. 161. I shall here content myself by repeating its principal characters, and adding some original observations.

This gypsum is found on a ridge of rock of about 2,400 metres [7874 feet] elevation; it does not itself form the ridge, which is calcareous, but is a little below. The beds are nearly horizontal. The gypsum is worked to the depth of about two-thirds of a metre; but the floor of the bed has not been discovered, being concealed by the debris of which this slope is covered; the roof is a limestone that is a little crystalline, blueish grey, and very schistose from a mixture of talc. This rock has not a thickness up to the surface of more than a metre or metre and a half [about 3 feet 3 inches to 5 feet]; it is full of crevices.

The workings cover a space of from six to seven metres [about 19½ to 23 ft.] in length, and as the gypsum is quarried solely for the use of the inhabitants of the valley, it is rarely worked. Mounting the ridge of rock, debris are only found, and no out-crop of the gypsum, and the principal inhabitants have assured me that there did not elsewhere exist the slightest trace of it throughout the valley.

Among the debris covering the slope much schistose limestone is met with, and also numerous fragments of quartz slightly micaceous. Notwithstanding the singularity of this position, it would be difficult to refuse admitting the contemporaneous origin of the gypsum and limestone; but I shall remark:

1. That this schistose limestone bears, from its colour, its mixture of talc, and all its other characters, a great relation to that so abundant in the Tarentaise, and also to that covering the gypsum at Brigg.

Those accustomed to the study of rocks know, that there exists between the members of the same formation an assemblage of characters that the eye often seizes at the first coup d'œil, but that are extremely difficult to define.

2. That all these fragments of quartz, often large and always angular, can only arise from superior rocks, there is every reason to suppose that the schistose limestone of the roof contains quartzose veins, as also happens in transition limestone.

3. Lastly, I will call to mind the mineralogical relations which unite this gypsum to all the others I have described, and above all to that of Brigg, of St. Leonard, and others that belong to the transition series.

To all these considerations, that already lead me to refer this gypsum to the same era of formation as those that I have noticed, I shall add another; it is, that the strong differences of nature and structure, which I have observed in this valley of Cogne, between the rocks of the heights and those of the valley, lead me strongly to suspect that the lower part is a primitive country crowned towards the summits, or at least the lower crests, by transition rocks; but I could not sufficiently assure myself of this fact to venture to give it as certain.

7. *Gypsum of the Val Canaria.*

This small valley, but two leagues in length, descends nearly from N.E. to S.W. into the high valley in which the Tesino flows; its opening, which is very narrow, is at a short distance from the village of Airolo, at the foot of St. Gothard. Following the torrent at first, we find ourselves between enormous masses, and as it were between walls of the micaceous gypsum I have above described. Its stratification is not very distinct, and varies greatly. I observed beds with all kinds of directions and inclinations; in some places they were horizontal, some were contorted.

This first observation struck me, by bringing to my recollection similar irregularities in the gypsum of the Tarentaise, and many others.

Large angular blocks of mica slate are seen on the banks of the torrent, some mixed with garnets, and others with hornblende. I took the greatest pains in searching for the beds of mica slate, which were stated to be associated with the gypsum, but I could not discover the slightest trace of it. The gypsum is there, as in many other places, absolutely isolated.

It was nevertheless possible that this association of gypsum with mica slate occurred towards the top of the slopes, and as it was announced (Itinéraire du St. Gothard, p. 73.) that the gypsum continued on the south towards the Val Piora, I preferred ascending the mountain on my right, that is to say, on the left bank of the torrent. I had observed this mountain from Airolo; I had remarked that its rocks were most frequently naked, and that their highly inclined beds appeared in their direction to cut those of the valley; I was then assured, that in crossing these beds, I should meet with gypsum, if it were true that it was subordinate to the mica slate.

After half an hour of a very steep ascent, always on gypsum, I arrived at a more even surface, only slightly inclined towards the valley; ascending the valley, I followed for some time this species of plain, and observed on its surface a great many of those holes and funnels so common in the gypsum rocks.* I soon turned and continued to ascend the slope, but I almost immediately found that I quitted the gypsum; and when I was a little more elevated so as to be able to command the bottom of the valley, I saw clearly that the gypsum did not extend very far along its bottom, but that it was no where elevated above the height I had quitted; that it only filled up the bottom of the valley, resting on both slopes, and that the small valley, or rather inferior ravine, was hollowed out of its mass. I assured myself the next day that the structure on the other side was similar.

* The like exist in the gypsum deposit of the Tarentaise, near the lake of Tines.

I had already every reason to suppose that the gypsum was of a formation posterior, both to the rocks of the two sides, and to the opening of the valley of which it appeared to fill the bottom.

I continued to ascend, and after having traversed the woods, where some projecting rocks shewed me mica slate but never gypsum, I arrived at the inclined high pastures, from whence I gained the foot of the upper escarpments of mica slate.

The stratification was perfectly regular; the beds have a direction nearly E. and W. and dip at 50° towards the N.; that is towards the commencement of the valley: it appeared to be the same on the opposite mountains; of this I have since assured myself.

From the inclination and direction of the beds, it was evident, that if the gypsum below was of the same formation as the mica slate, I ought, while crossing all the edges on the summits, to meet with at least some one of gypsum.

For this purpose, I began by retrograding to the anterior part of the mountain, towards the leventine valley, and ascending it, traversed for three hours the whole of the crest. It was in vain that I searched for gypsum; there was not the slightest trace of it. I once thought I had met with it, observing from a distance a bed of tolerably pure white; but on visiting it, I found it was formed of a decomposed dolomite.*

I have some reason to suppose that this bed of white dolomite, which is seen from a distance, has been taken for gypsum. The annexed design (Plate 2), representing the plan and sections of the valley, the disposition of the mica slate beds, and that of the mass of gypsum, will I think shew, in an evident manner, the impossibility of admitting the identical formation of these two rocks.

It is possible that some rock may have been found subordinate to the gypsum; but after consulting all the authors

* It dissolves in heated nitric acid, and leaves a residue of plates of talc. I have even determined by means of a rigorous analysis, the total absence of gypsum.

that have spoken of this gypsum, I have not been able to find any direct or positive observation. M. Lardy, an able mineralogist, a notice by whom is seen in a preceding part of these Annales,* on the gypsum of the Val Canaria, does not mention any foreign bed in it; he admits, as I have above said, that no gypsum is met with at a certain height; which is the more remarkable as he joins in the received opinion of the gypsum being primitive; but he rests this only on the relations that he thought he saw between the direction and inclination of the gypsum beds and those of the mica slate beds, relations which I am far from admitting, on account of the irregularity noticed above in the stratification of the gypsum. He concludes from these relations that the gypsum there forms a thick bed under the mica slate.

It is impossible for me to admit this opinion; it appears to me on the contrary, evident, that if the gypsum occupying all the lower part of the valley was included in the mica slate, or was covered by it, that it must, from the disposition of the mica slate beds and their escarpments, be found in some part of the heights; and it is from its total disappearance, beyond a certain level, that I found my conclusion that the gypsum is not contemporaneous, but posterior to the mica slate, and that there is no reason for considering it primitive.

It now remains to determine to what formation it belongs. But as it is not covered by any other rock, it is impossible to decide this question in a positive manner: I shall hereafter throw out a conjecture that appears to me very probable.

I shall here only add that the formation of the gypsum appears to be later even than the hollowing out of the Val Canaria. The contraction of that valley at its mouth, and the form of the gypsum mass, which affords a tolerably even surface, present more than in other places the idea of a deposit in a basin, or as has been said in a lake; an idea

* I have not translated this paper, M. Brochant's description rendering it unnecessary. (Trans.)

that I do not pretend to generalise, but which should not apparently be entirely rejected, at least with regard to a part of the gypsums of the Alps.

8. *Other localities of Gypsum.*

I have confined myself to describing the gypsums that have fallen under my own observation. Having been occupied so many years with this kind of rock, it may be imagined that I have consulted several works containing details on its mode of occurrence. I might therefore multiply quotations, bring forward numerous facts, and, by discussing them, add new proofs in support of the opinions I have already expressed in the examination of each locality observed; but I do not wish to extend the limits of this memoir.

I shall content myself by observing that all the other facts that I have been able to collect with regard to the gypsums of the Alps, (at least between the Mont Cenis and St. Gothard) may be arranged under three heads:

1. Those that are at the surface, and which do not appear to contain any bed of a foreign substance. The greatest number are of this kind; from this circumstance has often arisen the idea that these gypsums were the remains of ancient deposits in narrow basins.

2. Those which alternate with well characterised transition rocks, principally with argillaceous slates.

3. Lastly, those that are considered as primitive. I have seen but one example of it quoted (besides those of Cogne and St. Gothard, which I have described above); it is in the Haut Valais, near Lachs. I have not been able to observe it, having had no knowledge of it before my last tour.

From what I know relatively to this last, it does not appear that it is from a decided alternation with primitive rocks that it has been referred to that formation, but solely because it contains mica in tolerable quantity, and in sufficiently continuous veins; a character altogether insufficient, as I have shewn, since mica exists also in the same manner in admitted transition gypsums, and as this substance occurs in many other transition rocks.

Its primitive origin is then any thing but proved. I may even go further, for I strongly suspect that this gypsum is of the same epoch, and occurs in the same manner as the other gypsums of the Valais that I have before described; it would indeed be very extraordinary that the Valais, containing at least three portions of well recognised transition gypsum, should also present a primitive one, which is in other respects analogous to the former: and if such were the fact, we should have reason to be astonished at meeting but with this single example in the Alps, when nature exhibits herself so much in the great scale in that chain, and the same rocks are distributed over such a vast extent of country. I shall moreover add, that the Alps are the only chain in which primitive gypsum has been supposed to be recognized; it will then be the only primitive rock of which an analogous one has not been found in some other chains.

It remains for me to sum up the conclusions to which the facts I have produced have led me.

Summary.

1. It does not yet appear proved that there are in the Alps or elsewhere, beds or masses of gypsum distinctly contained in primitive rocks, and possessing characters of contemporaneous origin with them. (I have demonstrated above that the gypsum quoted as primitive in the valley of Canaria, and in that of Cogne, is of later origin.)

2. Many gypsums of the Alps form beds in a decided transition series.

The beds with which this gypsum is associated are limestone and argillaceous schist, which here represent the greywacke or anthracite formation. (The gypsum of Cogne, of Brigg, and of Bex, belongs to this class.*)

* Of these examples that of Bex, at least, is secondary. (Trans.)

3. There also exist in the Alps many deposits entirely superficial, most frequently in isolated masses of little extent; these gypsums most frequently rest on a formation of transition argillaceous slate or anthracite (as in the Tarentaise or Allée blanche); sometimes on transition limestone (as at Pesey and elsewhere in the Tarentaise; lastly, even on primitive rocks (as at the Val Canaria); many have an analogous appearance to a deposit in basins.

I am fully sensible, that it will be necessary to determine in a more precise manner the geological relations of those last gypsums that are met with on the surface.

But I confess that I am not in possession of sufficient data to resolve this question completely.

In the first place, these gypsums do not contain any foreign rock; at least I do not know of any; and they are not covered by others. They bear many mineralogical resemblances to the transition gypsums; they often rest on transition rocks; I am acquainted with a locality (near Bex, in the direction of the mountains on the N.E.) where they almost form a continuation of the decided transition gypsum, and for an extent of many leagues, except with some interruptions.

One would be led to imagine that these superficial gypsums were posterior to the former, but that they equally belonged to the transition series; that gypsum was deposited at different times during this period of formation, so that the first deposit was in the heart of the series, and the latter on its surface, without however shewing itself associated with rocks of later formation. But on the other hand, these same superficial gypsums of the Alps, bear many resemblances to the ancient secondary gypsums observed in Bavaria, Salzbourg, Thuringia, and elsewhere. The latter are also for the most part equally white; they contain like the former, anhydrous sulphate of lime, rock salt, and saline springs; they are frequently close to fetid limestone, which is not rare in the transition rocks of the Alps; in a word, there is but one difference between them, but that is in truth important; it is, that they rest on secondary rocks, and form part of a series essentially posterior to the transition class. *

* As the gypsum of Bex has been shewn to be secondary, and to form part of the saliferous or new red sandstone series, and as M. Brochant de Villiers mentions a locality near Bex, where the superficial gypsum and that of the latter place are seen to be almost continuous, there is very good reason for supposing that all the superficial gypsums of the Alps may be referred to the same class, and it is possible that many of the others also may belong to the same epoch, particularly when it is recollected how much confusion has existed relatively to the transition rocks of this chain; of this we have a striking instance in the rocks near Bex. The difference between these gypsums and those of Germany considered as important, viz. one class resting on primitive and transition, and the other on secondary rocks, can be of no consequence, any superior rock may rest upon an inferior one, the intermediate rocks being wanting: thus for instance, the oolite formation of the Jura rests on gneiss on the Rhine, at Lauffenburg, and lias on granite between Rouvray and Maison Neuve (department of the Côte d'Or, in France.) (Trans.)

Geological Sketch of the Thuringerwald. By M. de Hoff, Counsellor of Legation at Gotha.

(Extracted from M. Leonhard's Annuaire Mineralogique, by M. de Bonnard, Engineer in Chief of the Mines.)

(Annales des Mines for 1817.)

M. de Hoff has published a description, in two volumes, of the Thuringerwald, a small chain of mountains, having a direction from N.W. to S.E. and separating Franconia from the Duchies of Saxony, the whole of which is also known under the ancient name of Thuringia. The geological part of this work has been extracted in M. Leonhard's Annuaire Minéralogique for 1815; of that extract we are about to give the substance. We shall also extract some notices: 1st, of a memoir by the same author on the secondary limestone of the northern side of the Thuringerwald, inserted in the Annuaire Mineralogique of 1810; 2dly, of a letter from M. de Hoff to M. Leonhard, inserted in the same work for 1811, page 375, &c.

The most elevated parts of the Thuringerwald are formed of granite, porphyry, and slate; the granite principally forms the western extremity and all the southern slope from the crest. It sometimes rises as high as the summit, and appears here and there on the northern slope, or on the Thuringian side; but is most frequently covered by porphyry and other rocks. This granite, very uniform in the nature of its constituent parts, varies greatly in their proportion and size. The quartz in it often forms veins or geodes, sometimes ferruginous, which have given rise to different mining works, or rather researches. It also forms large masses that are worked, when pure, for the porcelain manufactories. A vein is worked in this granite at Heisenberg, near Ruhla, formed of quartz, fluor spar, sulphate of barytes, and hornstone, containing oxide and carbonate of copper, with copper pyrites. A part of this vein entirely consists of hornstone, containing abundant fragments of granite.*

The granite, properly so called, is surrounded by thick beds of rocks, presenting passages from granite into gneiss, or porphyry, sometimes by true gneiss and primitive trap (amphibolites). The mountain of Træhberg, between Weissenberg and Winterstein, is entirely formed of a rock composed of felspar and hornblende. All these formations are covered by mica slate.

The gneiss and mica slates contain considerable beds of greenstone (diabase), and of porphyry of a compact felspathic base, passing, the author says, into hornstone, very different from the porphyry in large masses, of which we shall presently speak.†

All these rocks are traversed by numerous veins of quartz and heavy spar, which were formerly worked, but are now abandoned. Mica slate is not found to the east of Kleinschmalkalde; but rocks of hornblende, felspar, and quartz, occur in every variety of mixture and grain, and constitute whole branches of mountains. A little beyond, the granite disappears under red sandstone (alter konglomerat); farther

* This fact is analogous to that of the hornstone vein of Carlsbad. (Journal des Mines, vol. 38, 341).

† An example is here seen of the most ancient porphyry formation admitted by Werner. It is to be observed that the porphyry occurs here equally in gneiss and mica slate, whilst in the Erzgebirge it has not been found in the latter rock. (Journal des Mines, vol. 38, p. 440.)

still, it is covered by porphyry; it is again found only near Zella and Shul, remarkable for its singular structure, for the large crystals of felspar it contains, and for its passages into different rocks.* It is accompanied by primitive greenstone, and covered by argillaceous schists and clay slate, which occupy the place of mica slate in the eastern half of the Thuringerwald. The schists are of a blackish or reddish ash grey, contain an abundance of quartzose portions, either as grains, veins, or beds, and very much approach the nature of mica slates. Some penetrated by quartzose particles, are worked as Whetstone slate.

Porphyry is, of all the primitive rocks, that most abundant in the Thuringerwald, and forms the most considerable branches of two-thirds of the length of the chain. Beginning with the west, that of the principal, or second formation, which we are at present considering, has generally an argillaceous base (base d' argilolite), either red, greyish red, or pearl grey, and contains very variably sized crystals of felspar and quartz. The paste is sometimes of a green colour, and the rock becomes apparently a hornstone porphyry, or passes into a porphyry with a greenstone or trap base. These porphyries constitute an assemblage of steep rocks, of frequently many hundred feet elevation; and afford picturesque points of view in almost all the valleys. The mountain of Rupberg is remarkable, from the prismatic recesses the porphyry presents, as has been observed by Anschuez.† Very beautiful green porphyries occur round the Schneekopf, and in the environs of Schwarzwald, Oberhof, &c. Porous porphyries, whose cavities are often lined with small quartz crystals, are worked as excellent millstones. Their paste moreover contains much quartz and felspar. A tolerably

* M. von Raumer regards the granite in the environs of Suhl, as well as the porphyry that accompanies it, as belonging to the Sienite formation, and as resting on the slates; he supports this opinion by his own observations, and those mentioned in the work of M. Heim, on the Thuringerwald. (Geognostische Fragmente, p. 37, &c.)

† Uber die Gebirge und Steinarten des Kurfurstlichen Sœchsischen Hennebergs, p. 114.

extended commerce arises from these works, the products of which are known under the name of Krahwinkel millstones.

Advancing towards the east, a change takes place in the nature of the porphyries, and they become generally of a trappean base. Yet further east, beyond a line drawn nearly from Gehren to Eisfeld, the porphyries entirely cease, and give place to slates.

The upper beds of the porphyritic rocks, those which are contiguous to the amygdaloïd (variolite) and red sandstones that cover them, contain, in an argillaceous mass of a reddish grey colour, besides the ordinary crystals of felspar and quartz, cavities from the size of a pea to that of half a foot in diameter, but which are generally of the same dimensions in the same district; the interior of them is either filled with chalcedony or hornstone, or furnished with a layer of that substance, lined inside by crystallized quartz, clear, or of a violet or reddish white colour. There are also sometimes crystals of carbonate of lime; sometimes the cavity is filled with powdery oxide of iron. The porphyry surrounding these nodules, is penetrated by silex, which gives it greater hardness than the rest of the mass; so that it longer resists decomposition, and numerous balls of it are found in the neighbouring rivulets. This rock is named ball porphyry, (porphyre à boules).

Amygdaloid almost always accompanies the porphyry, and especially the ball porphyry (porphyre à boules); its base is of trap or wacke. The nodules are either in the form of almonds, and filled with carbonate of lime or chalcedony, or rounded cavities, void or lined with green earth, (chlorite baldogée), calcareous spar, or crystallized quartz.

Transition limestone is entirely wanting in the western part of the Thuringerwald. It appears on the contrary on the east under the greywacke and schists that accompany it; it is quarried in the valley of Sorbiz, near Tœschnitz, and known by the name of Schwarzburg marble. It is generally of a dark colour, often altogether black, with red and brown spots, and numerous veins of white calcareous spar. It contains trochites, spiral univalves (terebræ), and a few undetermined bivalves; but there are no traces of corals observable in it.

Argillaceous transition slate forms a considerable mass of rock in the eastern part of the Thuringerwald, from Steinheide to Lehesten, where it constitutes the whole of the northern, and a great part of the southern side; it is in general black or grey, easily divides into thin and even plates, and is worked as roofing slate in numerous places, near Sonnenberg, Lehesten, &c. It contains beds of drawing slate, whetstone slate, and flinty slate. Nodules of pyrites and quartz are also found in it. A singular variety of slate occurs at Feldberg, near Sonnenberg; it divides into very slender separate pieces, possessing a certain degree of tenacious softness, which renders them fit to be employed as styles, or pencils for writing on slate. It is pretended that this variety is not found elsewhere.

Greywacke occurs solely on the east of the Thuringerwald, where it occasionally appears to alternate with slate. M. Heim regards it as situated beneath it. With few exceptions it occurs only on the south, or Franconian side, and especially in the valleys running towards the Kronach; its beds are sometimes as much as twelve feet thick; but they become thinner as they approach the slate, and then form greywacke slate. The greywacke is most frequently grey, sometimes whitish, blackish, reddish, yellow, or greenish. Its paste is the same as that of the argillaceous slates: it contains felspar, quartz, silvery mica, and a reddish rock analogous to porphyry; the whole in rounded grains, that are rarely of the size of a nut. Rocks are observable near Oberhasloch, in which the greywacke resembles a conglomerate.

These conglomerates, and micaceous or quartzose red sandstones (konglomerat rothes und graues liegende), which together compose the formation named red sandstone, considered the most ancient of the secondary formations, are the most extensively spread of all the rocks in the Thuringerwald. It is this rock that forms the commencement of the chain on the west, rising rapidly from beneath the newer secondary formations; it soon forms entire branches of mountains in the environs of Eisenach. It stretches along the chain, on the Thuringian side, for a considerable distance. It also occurs in many situations on the Franconian side; it especially fills the valleys descending from the primitive chain, and, in two places, even covers, in very thick beds, the summit of this chain. It contains near Tabarz veins of fluor spar, heavy spar, and iron ore. Important iron mines are worked in it near Friederichsrode. The rocks constituting this formation are of very various natures; sometimes formed of rolled fragments of the size of the head, sometimes their grain is that of the finest sandstone, sometimes they altogether resemble sandy or argillaceous schists. The intermediate varieties between these extremes are innumerable. When the rock does not contain any visible fragments, it is most frequently formed of red or green schistose clay; this same substance forms the paste elsewhere enveloping fragments, the nature of which is that of the surrounding primitive rocks. Sometimes it contains calcareous parts; it is often a fine sandstone, with parcels of mica, and containing nests or nodules of clay.

The coal formation of the Thuringerwald appears to belong, as a subordinate member, to the preceding, to which, according to the author, must be referred the sandstones and schistose clays accompanying the coal, and with which relations always exist, but in a manner difficult to determine. The coal formation occurs only in scattered patches, situated generally at the upper extremity or commencement of narrow valleys, filled by the red sandstone rock. The roof and floor of the coal are commonly of schists and sandstones, with impressions of ferns. At Tenneberg, near Tabarz, the coal is covered by Zechstein, or alpine limestone, a rock between which and the coal formation relations often also exist.

The alpine limestone occurs in the Thuringerwald, of very variable thickness, and of very different nature. M. de Hoff describes under this name: 1. marno-bituminous schist; 2. Zechstein; 3. gryphite limestone; 4. fetid limestone (stinktein); and 5. Rauhkalk.* The marno-bituminous schist, when it does exist, always constitutes the lower bed of this formation; is remarkable for the metallic

* A peculiar variety of alpine limestone is described under the name of Rauherkalk, or Rauhkalk, M. Voigt was the first thus to particularise and name it (Praktische Gebirgskunde, p. 87). Its fracture is granular, and it often appears penetrated by spathose particles; its texture is porous, full of cavities; these vary from the size of a pea to an inch in diameter; its name arises from the harshness of its touch. Dr. Jordan has given a new description of it in his Mineralogische Reise Bemerkungen, p. 57; he refers it to stinkstein, but believcs it different from that he has described under the same name of Rauhkalk, (Mineralogische und chemische Beobachtungen, p. 118), which exists on the south of the Hartz, and in which occurs the cavern of Einhornloch, near Schartzfeld. M. Haussmann describes this limestone of Schartzfeld, under the name of Blasiger Flœtzkalkstein, and considers it as above the ancient secondary gypsum, and consequently as posterior to the alpine limestone (zechstein) situated beneath the gypsum; he says nevertheless that there appears to be a passage from the stinkstein to the Blasiger flœtzkalkstein or rauhkalk of Jordan. (Nord-deutsche Beytrœge zur Berg und hüttenkunde, No. 2, p. 100.)

M. Freiesleben, in his interesting work on the copper slate of the Mansfeld country, describes, under the name of Rauchwacke, a brownish grey limestone, containing silex, sometimes bituminous, harsh to the touch, full of strange shaped cells, occurring immediately above the zechstein, that is, generally beneath the gypsum and stinkstein, and which he considers as different from the rauherkalk of M. Voigt.

The same name of Rauchwacke is given in the copper slate mines of Rielgsdorf, in Hesse, to a limestone also of a smoke grey colour, and very cellular, but which occurs above the zechstein, the ancient gypsum, the stinkstein, and even, as it appears, the second gypsum, and the clay accompanying it. This rauchwacke is then very different, geologically speaking, from that of M. Freiesleben.

We have thought it right to place these notices together, the discussion of which would produce long developments, in order to call attention to the confusion that exists relative to the names of Rauhkalk, Rauherkalk, and Rauchwacke, and to prevent the errors that it may occasion.

Note by the Translator.—In order to shew the analogies of the Alpine limestone, &c. to the magnesian limestone, and new red sandstone of the English series, I shall present the classification of the former series ores it often contains disseminated through its mass, (giving rise to numerous works), for numerous impressions of fish, and more rarely for the impressions of plants it affords; lastly, adds M. de Hoff, but without noticing the locality, for the beds of coal it contains. Near Kupfersuhl, at the western extremity of the chain, the impression of a skeleton has been found in it, which has been recognised as belonging

by Werner, and Freiesleben, as given by Heron de Villefosse (Richesse Minerale, vol. 2. p. 396, &c.), adding their equivalents in the English series.

CLASSIFICATION OF WERNER.

Oolites and Lias......	1. Newer limestone (neueres kalkstein) with which is arranged the Cavernous limestone (böhlen kalkstein) of the Jura, Swabia, and Franconia.
New Red Sandstone ..	2. Red clay with gypsum; 3. Variegated sandstone (bunter sandstein), with oolites; 4. Formation of rock salt and clay; 5. Ancient gypsum and fetid limestone;
Magnesian Limestone .	6. Ancient limestone, comprising: Cellular limestone (rauchwacke) Friable marl (asche) Gryphite limestone (gryphitenkalk) Compact marly limestone (zechstein) Marno-bituminous limestone, named by some bituminæser mergelschiefer, by others kupferschiefer, (copper slate);
New red Conglomerate	7. Floor of ancient sandstone (rothe todte Liegende.)

CLASSIFICATION OF FREIESLEBEN.

Oolite and Lias	1. Shell limestone (muschel kalkstein).
New red Sandstone...	2. Formation of sand and clay (sand und thon gebirge) comprising, without any determined order of superposition: Clay (named letten) Marl (mergel) Gypsum (thon-gyps) Sandstone (sandstein) Argillaceous iron ore (thonartigea eisenstein) The coal of this epoch (steinkohle) Limestone (kalkstein) Oolite (roogenstein) Sandy schist (sandschiefer);

to a species of crocodile, and which is now at the royal collection of Berlin. Proceeding eastward, the marno-bituminous schist is found on the northern slope, as far as Saalfeld; on the Franconian side, only to Flohe and Goldlauter. The metallic richness of this schist is very variable; it most frequently contains copper pyrites, sometimes grey copper, and carbonate of copper, at others only sulphuret of iron; in some places galena and little or no copper; in others, it contains no ores at all.

The marno-bituminous schist is generally covered by zechstein, which often contains gryphites, more rarely corals, and very rarely the points of belemnites. Gryphites are met with also near Schmerbach, with terebratulæ, in the upper laminæ of the schist bed, the inferior part of which contains the impressions of fish; sometimes even the same laminæ contain the gryphites and the impressions. If the latter, as has been said, are freshwater fish, the presence of

Magnesian limestone..	3. Ancient limestone (älterer kalkstein) containing: a. Gypsum and fetid limestone (gyps, stinkstein) b. Fetid limestone (stinkstein) with Calcareous iron ore (kalkeisenstein) Friable marl (asche) Limestone, known by the various names of böhlenkalkstein, rauhkalk, and rauchwacke; c. Compact marly limestone (zechstein) d. Marly schist (mergelschiefer) The roof of the copper slate (dach) The bed of copper slate (kupferschiefer flœtz) The immediate floor of greyish white sandstone (weiss liegende);
New red Conglomerate	4. Ancient sandstone (alteres sandstein gebirge) comprising: The inferior floor of red sandstone (rothe liegende) called floor in respect to the copper slate;
Coal measures.........	The coal measures (steinkohlen gebirge).

For further information respecting the connexion of the alpine and magnesian limestones, with full details of the latter, I refer the reader to that excellent work, Messrs. Conybeare and Phillips' Outlines of the Geology of England and Wales.

marine fossils in the same bed is rather a remarkable fact. The remains of fish do not exist every where in the marno-bituminous schist. They occur near Ilmenau in a species of flattened nodules, situated in the middle of the bed; upon clearing the nodules, the two impressions are seen, which are sometimes entirely filled by calcareous spar.

The well characterised zechstein, of a smoke grey colour, slightly slaty, with a splintery and almost uneven fracture, is generally but few fathoms thick. When it encreases much in mass, it changes its nature, and its upper parts appear as stinkstein, or rauhkalk. As soon as it acquires the porous and cavernous texture of the last variety, it becomes of enormous thickness, and forms rocks and entire mountains of six hundred feet (English) and upwards in height. It is difficult to assign the constant relations between stinkstein and rauhkalk, which often pass into each other. Yet the latter most frequently covers the stinkstein, the smell of which, sometimes scarcely perceptible, appears owing to sulphuretted hydrogen gas, and not to bitumen. Both the one and the other do not appear to contain fossils. Many considerable caverns are known in it, among others one situated near Altenstein, from which flows a considerable stream, not far from Glücksbrunn.

The red sandstone is sometimes wanting on the south of the Thuringerwald, and the alpine limestone rests immediately on granite; on the north, the superposition of the limestone on the red sandstone, forms altogether a straight line, prolonged without deviation, for many leagues across mountains and valleys. It probably results from this: 1. that when the sea deposited this limestone, the shore, formed of red sandstone, was thus cut, it may be said, in a straight line; 2. that at this epoch the present valleys were not formed, or did not descend below the surface of the sea, since the waters would have entered into these valleys, and would have deposited the limestone, in the same manner as the red sandstone and coal formation have been deposited in the first valleys of the primitive rocks; 3. that the waters under which the red sandstone and coal formations were deposited, were much more elevated than those which afterwards formed the limestone beds. The author draws different conclusions from it on the successive formations of valleys; but confesses that many facts appear to him inexplicable.

M. de Hoff says that the disposition of the gypsum, sandstone, and shelly limestone, more recent than alpine limestone, and surrounding the Thuringerwald, is known from the writings of M. Voigt. We shall only notice the superpositions made known by a gallery, named Herzog-Ernts-Stollen, pierced into the northern slope of the chain, near Friederichroda, in order to discover the disposition of the beds, and to give occupation to poor miners.

This gallery is pierced, 1, in variegated sandstone (bunter sandstein) very thick beds of which it at first traverses, they contain subordinate beds of clay, which are often six or twelve feet thick. Afterwards occur, 2, four fathoms of a fine grained limestone, disposed in beds from a few inches to a foot in thickness, with a regular direction towards about NW. by W. (9 or 10 heures), and inclining towards the east. This limestone is of a clear grey, or of a yellowish grey, granular, or almost spathose, often full of small salient points of a darker colour, which appear with a microscope to be spathose grains, and give the rock the appearance of a fine grained oolite. No fossils are seen in it, and it gives out no smell. It is cut by a nearly horizontal vein of red clay, a foot thick, without the beds being at all shifted; but on the wall of this vein, the calcareous beds have a spongy exterior, about an inch thick, containing large spathose spots. Beneath this limestone, the gallery traverses; 3dly, a bed of clay; 4thly, a soft marly limestone of a whitish grey colour; 5thly, another clay bed; 6thly, another bed of grey limestone, compact, and a little cellular; 7thly, clay; 8thly, a thick gypsum bed, containing cavities, in which superb crystals of selenite have been found half a foot in length. It is only in this place, and near Kittelsthal, that ancient secondary gypsum has been recognised on the north of the Thuringerwald. 9thly, grey clay is afterwards met with;

then, 10thly, zechstein, with its ordinary characters; 11thly, marno-bituminous schist; 12thly, red sandstone (todliegende); and lastly, porphyry, into which the gallery has only been driven a few feet. The total length of this gallery is 430 paces.*

Basalt occurs only at some distance from the true chain of the Thuringerwald, and forms some insulated mountains. These are the Pflasterkaute, near Eisenach; the Dollmar, near Kühndorf; and the Steinsburg, near Suhl.

* We again call the attention of our readers to M. Freiesleben's work, entitled, " Geognostischer Beytrag zur kenntniss des Kupferschiefergebirges, printed at Freyberg, in four parts, 1807 and 1816," and in which the succession, the nature, and mode of existence of the different secondary formations of the country of Mansfeld and Thuringia, are exposed and developed by numerous and instructive details. A notice of these different formations has been given by the Count Dunin Borkowsky, in the Journal des Mines, vol. 26, p. 182. (Note, Annales des Mines.)

On some basaltic mountains of Hesse and Thuringia. By M. DE HOFF.

Extracted by M. de Bonnard.

(Annales des Mines for 1817.)

A Memoir by M. de Hoff, inserted in the 5th volume of the Magasin der gesellschaft naturforschender freunde (Berlin, 1811), contains a detailed account of many basaltic mountains, noticed slightly at the end of the preceding paper.

The author in the first place remarks that the southern part of Germany is rich in basaltic rocks, which are on the contrary wanting on the north. He adds, that if a line is drawn from Eisenach to Münden, and it is prolonged towards the N.W., and another line from Haute-Lusace by the Erzgebirge, cutting the first at an angle of from 95° to 100°, near Culmbach, in the country of Bayreuth, all that part of the European continent, situated to the north of these two lines, does not contain any basaltic rock,* which

* M. Haussman however mentions basalt near Sandefiord, in Norway; but considers that he ought to refer it to the transition series, as well as all the trapean and quartzose rocks that accompany it. He adds, that the south of Norway affords numerous geological paradoxes. Are the rocks mentioned by M. Haussman really basalts, in the general acceptation given to that name? (Note of the editor of the Annales des Mines.)

gives a peculiar interest to the mountains of this nature that exist on its limit; such are the Meisner, the Blauekuppe, near Eschwege, the Stoffelskuppe, and Pflasterkaute, between Eisenach and Marksuhl, the Steinsburg near Suhl, the Dollmar near Meinungen, and some other points situated between these, which are all on the first of the lines noticed at the limits. These mountains are moreover isolated in the midst of secondary rocks; their position is entirely independent of the chains of mountains and principal valleys in the neighbourhood, and it is at some leagues from them that the extended basaltic rocks of Hesse and Franconia commence. For that which relates to the Meisner, M. de Hoff refers to the works of Messrs Voight and Schaub.

In the high mountains mentioned and described by M. de Hoff, this remarkable fact is observed, the basalt does not cover, as at the Meisner, the secondary rocks with which it is in contact; but, on the contrary, penetrates through them, forming as it were wedges, that enlarge in proportion as they descend. This fact had already been noticed by M. de Voight, in his Mineralogische Reise, and in his Kleine mineralogische Schrifte, and by Messrs Sartorius and Gœrwitz, in the work entitled Die Basalte in der gegend von Eisenach; but the new observations that M. de Hoff has been enabled to make, in the portions of basalt recently laid open by quarries, and the plates with which his memoir is accompanied, can no longer leave any doubt on this head.

The mountain named Blauekuppe, situated one league to the south of Eschwege, is isolated, of a form a little elongated from S.W. to N.E., and composed, as is the surrounding country, by reddish or yellowish sandstone, in horizontal beds from a decimetre to a metre [about 4 inches to 3 feet 3 inches] in thickness. On the summit of this mountain, and in the direction of its length, a projecting crest is observed, formed by a basaltic mass, which cuts down through the sandstone beds, as may be proved in the quarry situated to the south-west.

At a distance from the basalt, the sandstone is in reddish or whitish thin beds; it contains numerous small plates of mica, flattened nodules of clay are observable in it, which belong to the variegated sandstone formation* (bunter sandstein.) Approaching the basalt, the sandstone becomes yellowish, loses its granular texture, and acquires an earthy and dull appearance. The horizontal fissures disappear, and are replaced by beds of brown clay some inches thick; and a number of clefts and fissures are seen in it.

The wedge of basalt, which is only from three to four metres [about 10 to 13 feet] thick at the summit of the mountain, becomes thicker as it descends. On the east side it is separated from the sandstone by an open cleft, as much as a decimetre [about 4 in.] wide, and the sandstone is, on the side of the cleft, covered by a vertical lining, from five to six millimetres [about $\frac{1}{4}$ in.] thick, formed of an argillaceous sand of a greenish yellow, penetrated by dendrites. Towards the west the basalt is in immediate contact with the sandstone. An open and almost vertical cleft is observable in the interior of the basaltic mass, dividing it into two unequal parts, the nature of which is not the same; there are moreover a multitude of little fissures irregularly disposed. Towards its western end, the basaltic mass forms a hook, or almost horizontal branch, which penetrates some metres into the sandstone. The fissures of this portion are almost all arched, and nearly parallel to the curve forming the hook.

The basaltic mass contains, in its upper part, the remains of sandstone beds disposed horizontally, and divided also by the vertical cleft of which we have spoken above. It contains moreover not only common basalt, but, 1st, basalt in balls, in an aggregate that in some respects resembles greywacke, which is penetrated by calcareous veins, and contains small crystals of oxidulated iron; 2ndly, a basaltic amygdaloid (variolite), containing cavities filled by the usual substances in such cases, but also enclosing fragments of sandstone; 3dly, the basaltic hornstone of Messrs. Sartorius and Gorwitz; 4thly, the lave lithoïde petrosiliceuse of Dolomieu; 5thly, the blistered lava (lave boursouflee) of

* New red sandstone formation. (Trans.)

the same. M. de Hoff conceives that the two latter substances cannot be confounded with any of those named variolites or amygdaloids, and that they are very evidently lavas.

Basalt, properly so called, constitutes especially the portion of the mass comprised between the middle cleft and the eastern part (also separated from the sandstone by a cleft); the variolites and the blistered lavas form on the contrary the western portion, and especially the hook that descends into the sandstone.

The nature of the sandstone and its change in the neighbourhood of the basalt, the fragments of sandstone beds that the latter contains in its upper part, the form of the basaltic mass, the hooked form it presents on the west, and by which it appears to have sought to escape, before it was able to raise the upper crust of the mountain; lastly, the true lavas with which the basalt is associated, appear to M. de Hoff sufficient reason for believing that the wide cleft filled by basaltic rocks has not been filled by precipitation, and from above, but from beneath, and by the raising of a mass in a liquid state. He considers therefore that this situation speaks in favour of the volcanic origin of basalt.

To the S. E. of the Blauekuppe, 1st, between Stedtfeld and Horsel, in the country of Eisenach; 2dly, near Hütschof, on the road from Eisenach to Berka; 3dly, at the Kupfergrube, between Weinschensuhl and Horschlitt, three facts are observable analogous to those seen at the Blauekuppe, that is to say, the basalt traverses beds of variegated sandstone.* The same fact is also remarked at the Stoffelskuppe, situated at a greater distance in the same direction; it has been described by M. Danz, counsellor of mines, in the second volume of Observations of the Society of Berlin. Fragments of sandstone are also seen in the basalt of this place.

The quarry of Pflasterkaute, near Marksuhl, is but half a league from the Stoffelskuppe, and the basalt of the one is probably the continuation of the other. The basaltic mass

* New red sandstone. (Trans.)

is here from three to four times thicker in its upper part than at the Blauekuppe; it in the same manner traverses almost vertically the horizontal beds of sandstone. It is here also seen to become thicker as it descends, and in the western part of the quarry a considerable mass of basalt is seen altogether beneath the sandstone, the beds of which surround it in a singular manner, following all the sinuosities of the upper surface of the basalt, an effect that could only arise in consequence of the softening of the sandstone beds, occasioned by the heat of the basaltic mass in fusion.

The nature of the sandstone is also altered in the neighbourhood of the basalt; it becomes very divisible, earthy, of a brown colour, full of dendrites, and offers all the characters of that of the Blauekuppe. The basaltic mass is composed in part of basalt, containing zeolites, chalcedony, olivine, calcareous spar, and oxidulated iron, and in part of variolite, and basaltic hornstone (hornstein basaltique); hollow nodules are met with in it, of which the crust is formed of oxidulated iron, and the interior lined by calcareous spar in dodecahedral crystals.

The Steinsburg, near Suhl, is a mountain formed of nearly horizontal beds of variegated sandstone (bunter sandstein.) A basaltic ridge appears on the summit 20 metres [about 66 feet] thick. This ridge appears on the surface, for a length of about 120 metres [393 feet], in a direction from S.W. to N.E. On the slope of the hill is a quarry in the sandstone. A society of geological amateurs united in order to have a gallery pierced from the quarry to the basaltic mass, at about 20 metres [about 66 feet] from the surface, and at eight metres [about 26 feet] in vertical depth, the basalt was met with still cutting and traversing all the sandstone beds, as it descends. The basalt is separated from the sandstone by a species of vertical crust of sandstone about two centimetres [nearly 1 in.] thick, afterwards by a bed of soft clay, of a blackish grey colour, three centimetres [a little more than 1 in.] thick, in which are fragments of sandstone, and which also contains tables (tables) of basalt; the basalt is afterwards found in tables disposed parallel to the side of the mass; and lastly the basaltic mass, full of irregularly disseminated clefts. This basalt contains much olivine, hornblende, and felspar; it also contains fragments of sandstone, but neither variolites or lavas have been observed in it. As these relations of position to the sandstone are only made known by this gallery, it might be said that the basalt is here a vein in the sandstone; but M. de Hoff observes that every thing leads us to the belief, judging at least from analogy, that this position is of the same nature as those before noticed.

* The basaltic rocks mentioned in this paper are evidently the same as the basaltic and other trap dykes of the British islands. A very interesting and detailed account of the trap dykes of Anglesea by Mr. Henslow, occurs in the 1st volume of the Cambridge Philosophical Transactions, page 401, &c.; and those of the Hebrides, &c. will be found ably discussed by Dr. Mac Culloch in his account of the Western Islands, and in the Geological Transactions, which also contain many other descriptions of similar facts. (Trans.)

Report on the Tin of Piriac (department of the Loire Inferieure), by Messrs JUNKER *and* DUFRENOY, *Assistants to the Royal Mining Corps.**

(Annales des Mines *for* 1819.)

THE rocks of the coast from St. Nazaire to the mouth of the Vilaine, comprise two distinct but primitive formations. The first entirely granitic, extends from St. Nazaire to about two kilometres [8000 feet] on the S.S.W. of Piriac.

The second, which is schistose, forms nearly the whole of the remainder of the coast. It is at the separation of these two formations, that is to say, at the northern extremity of the granite, that the oxide of tin is met with.

Granitic Group.

The granite of St. Nazaire, Guerande, and Croisic, is generally small grained, of a brownish grey colour, with black mica, often traversed in every direction by numerous veins of fetid quartz. It sometimes contains crystallized tourmaline (schorl?) as also felspar. In some situations (at Turballe and Clis), the latter is found alone with the quartz, producing the mineral named graphic granite, which occurs in the most ancient rocks.

* I have extracted only that part of this Report which gives a geological sketch of the country in which the tin occurs. (Trans.)

The granite sometimes presents the appearance of stratification; the beds seem to have a direction from S.E. to N.W. and to dip towards the sea, i. e. towards the S.W. at a small angle with the horizon. It contains but few subordinate beds; some stripes of no great thickness are seen in it of very micaceous slate, or rather of mica in mass, and very large grained granite, especially at Clis and Pouliguen.

The part of this coast comprised between the Croisic to about 1500 metres [4839 feet] on the S. S. W. of Piriac, is still granitic, and presents nearly the same characters as the preceding, for two-thirds of the distance, i. e. veins are seen in it of quartz, of quartz and tourmaline (schorl?) of quartz and felspar, of graphic granite, and very fine grained granite.

As the point that we have mentioned is approached, at the separation of the two primitive groups, the grain of the granite becomes larger, particularly in the last 400 metres [1312 feet] which contain the oxide of tin. The slightest trace of stratification is no longer observed in it; the veins become thicker and more numerous, and scarcely contain any thing but crystalline quartz. All this extent of coast is covered by a bed more or less thick of quartzose sand; a species of dune or arid bank of many inches in height formed of this same sand, accumulated by the winds and waves, borders the sea, and serves as a dyke against it. Thus it is only on some masses, uncovered by the sea at low water, that we have been able to make our observations.

At a short distance from the separation of the two formations, the sandy dune begins to rise; farther on it exposes the granitic base to view, which it before masked; it is soon reduced to a few decimetres thick in sand, covering the rock that becomes more precipitous, and rises above a sandy beach successively washed by the tides.

Schistose Group.

The granite ceases near the hamlet named Penhareng, and is replaced by a suite of remarkable schistose rocks.

The surface at the separation of the two rocks, at least the part that can be observed on the coast, is tolerably regular, with the exception of two points which the granite forms in the schist, and round which the latter mantles; it has a direction from N.W. to S.E.

The first rock of this group, composed of mica and quartz in very variable proportions, occupies a space of about 50 metres [164 feet] in the parts nearest the granite; mica of a golden yellow colour is most abundant, the rock is very soft and decomposed.

Beyond this, the quartz augments and communicates to the composition its hardness and indestructible character. It is then in some places reddish, in others yellow; sometimes even the quartz is the sole ingredient, apparently without mica, and without change in the schistose structure of the rock; it is on a small scale, only very much contorted, and contains within its folds nodules of quartz.

The rock that immediately follows the last is composed of mica and tourmaline (schorl?) in very thin needles, crossing each other in every direction, and of a little quartz. It is not contorted, and contains veins of quartz and felspar parallel to the laminæ.

The tourmaline (schorl?) is sometimes so abundant in it, as to form black stripes or veins, giving it a ribbon-like structure.

The third rock, forming the point named the Bichet, is composed of grains of quartz, and crystallized felspar disseminated in a paste of greenish-grey mica; it sometimes contains in the direction of the laminæ veins of quartz, at others of felspar, often swelling out considerably.

The fourth rock differs from the preceding in no longer containing the grains of quartz and felspar; the paste alone remains, which is of a greenish-grey colour, very soft and unctuous to the touch.

It is covered by a singular rock, which is a quartz, possessing an imperfectly schistose structure, of a dark black colour, containing between its laminæ a black brilliant substance that strongly soils the fingers, and is probably combustible. This rock forms rather large veins much contorted and interlaced by veins of yellow quartz, than beds properly so called.

Very thin veins of yellow oxide of iron have been found in it, parallel to the laminæ.

The sea has hollowed a very deep cavern out of this rock, called Madame.

An extremely schistose rock, composed of greenish-white mica, grains of quartz and felspar, and sometimes crystallized garnet, afterwards occupies all that part of the coast comprized between the cavern of Madame and Castelli point, where the shore, after having attained its greatest height, 10 or 12 metres [about 33 to 39 feet 6 in.], begins again to get progressively lower. It is followed by the rock No. 4, then by the same black quartzose rock that we have just described.

Lastly, beyond Castelli point, near a cavern named Trou du Moine fou, small grained granite is found, resembling that which precedes the schistose group, but of little extent; it ceases in the midst of the port of Piriac, near which it participates in the schistose structure of the remainder of the formation.

This granite is covered by a greenish-grey schist, at first micaceous (composed of quartz and mica), but which soon changes its nature, containing hornblende and passing into greenstone.

The first of these schists contains between its laminæ some spots of sulphuret of iron, and constitutes the surface of the coast from the port of Piriac to the environs of Port au Loup. The second contains garnets as an accidental mixture, often in considerable quantity, and oxidulated iron, and extends from Port au Loup to the passage of Trehiguier on the Vilaine.

All these beds have a constant direction from S.E. to N.W. nearly, presenting a remarkable disposition with regard to their dip. Those which on the north follow the first stanniferous granite, to the middle of the greenish rock No. 6, situated between the points of Bichet and Castelli, dip towards the N.E. at an angle of from 45° to 50° with the perpendicular, whilst the others dip to the S.W. at an angle of 60° to 80°, and cover the second granite of Trou du Moine fou. The latter is above the schist of Port Piriac, which dips under it towards the S.W. at an angle of from 40° to 60°, and has a direction from S.W. to N.E. which is that of almost all the rocks of Brittany. It then appears evident; 1st, that the two groups we have described belong to two different formations; 2dly, that they are essentially primitive, and that the first, containing indications of tin, is the most ancient.*

Oxide of Tin in place.

Indications of tin are met with in the 400 last metres [about 1312 feet] of the granitic rock, and at about 1500 metres [4839 feet] to the S. S. W. of Piriac; the rocks or the veins they contain are only exposed in the numerous reefs projecting into the sea, which with the exception of some points or summits, are covered by every tide. The remainder of the beach to the escarpment terminating it, and against which the sea breaks in bad weather, is commonly

* This stanniferous district of Brittany, is most probably a continuation of the stanniferous rocks of Cornwall, more particularly of the Land's End; it does not appear however that the schistose group contains tin as in Cornwall; from a memoir published in the 2d vol. of the Geological Tranasctions of Cornwall, by Dr. Forbes, it appears that the Land's End district may also be divided into two groups, the granitic and schistose, the granite like that of Piriac, is stanniferous, and the schistose rocks may probably be referred to the same age as those covering the granite in that district; greenstone occurs in the schistose groups of both countries; the schorl (tourmaline) rock of Brittany appears as separated from the granite by a rock of mica and quartz, whereas in Cornwall it appears to be a modification of granite. With regard to the other schistose rocks, felspar and hornblende appear most abundant in Cornwall, and quartz in Brittany.

For a particular account of the schorl rock of Devon and Cornwall, consult Mr. J. J. Conybeare's Notice on the Geology of those counties, (Annals of Philosophy, vol. 5. new series, p. 188 and 189.) (Trans).

covered by a thick bed of sand, which, in many places, is interrupted by masses of granite or salient veins of quartz, harder and tougher than the rest, and which have resisted the effects of the waves and decomposition.

About the equinoxes, and especially that of Autumn, it sometimes happens that the sea removes this bed of sand and gravel, and when retiring shews the structure and mode of occurrence of the veins that traverse the ground in every direction. But shortly (generally at the next tide) the same sands are brought back again ; and the inhabitants of Piriac are the only persons who can profit of these valuable moments, in order to observe the veins, and collect without trouble, the fragments of ore broken off from their places, and washed by the sea.

[In order to ascertain the thickness of the sand and to study the rock beneath, the authors opened a trench 200 metres [about 656 feet] long parallel to the west, and extending to the separation of the granitic and schistose rocks].

The sand covering the rock being about three or four decimetres [about 12 to 16 in.] thick, was soon traversed. We constantly found in the inequalities of the granite surface, that it covered tin sand, and pebbles having a mean diameter of two to three millimetres [about $\frac{1}{8}$ in.], many of which still shewed crystalline forms.

The granite occurring beneath is generally very soft, and the felspar in the state of Kaolin. From this circumstance it has been incorrectly stated that the oxide of tin was found in clay at Piriac. This granite and the rocky masses advancing into the sea, are traversed by numerous veins of greyish-white and fetid quartz, often containing crystals of felspar and mica. These veins are of very variable dimensions, being from one and two centimetres [about $\frac{1}{2}$ in.] to many metres in thickness. Their direction and inclination also varies, they cross each other in every direction ; often swell out considerably, forming with each other a true stockwerck, analogous to those of Geyer in Saxony, and affording a disposition resembling that of the carbonate of magnesia and common opal, (quartz résinite) in the serpentine of Piedmont.

It should be remarked that few of these veins are metalliferous, and that out of more than ten cut by the trench, two only afforded oxide of tin, and those irregularly and without continuity.

[Here the authors describe veins bent in different directions.]

The trench was cut for some metres in the quartzose mica slate covering the granite, but nothing was discovered in this part of it.

Many of the veins which cut the rocks covered by the tide, and especially those near the Tombeau d'Almanzor, having been found to be stanniferous, we attacked them with the intention of discovering the mode of occurrence of the ore, &c.

We found that the oxide of tin occurred in nests and veins, accompanied by quartz and mica, often decomposed. In nests when the quartz was abundant; in veins on the contrary, when mica predominated; in the latter case the tin is in greater quantity, and it is seen to augment the more the mica and granite are decomposed; the tin is then in a loose and friable state.

Considerations on the place that the granitic rocks of Mont Blanc and other central summits of the Alps ought to occupy in the order of anteriority of the primitive series. By M. Brochant de Villiers.

(Annales des Mines, for 1819.)

Read at the Royal Academy of Sciences, May 27, 1816.

The name of granite was formerly given to all mixed rocks composed of crystalline minerals. Saussure has often employed it in this general sense in his works.

The high Alps of Mont Blanc and St. Bernard, having been much more visited than other parts of the same chain, granites have been mentioned as occurring there, and have been cited in all works on geology.

During the last fifteen years, mineralogists have agreed to restrain the meaning of the word granite, and only to apply this name to rocks composed of felspar, quartz, and mica,* in a crystalline and not schistose state, and many other rocks formerly confounded with them, have been removed from them.

* Dr. Mac Culloch has observed (Classification of Rocks, p. 230 and 231), that "this distinction is too limited for practical purposes; and, in a geological sense in particular, it is inadmissible:" he states granite as consisting fundamentally of quartz, felspar, mica, and hornblende, variously combined, other minerals occasionally entering into its composition and forming integrant parts of it. For a synopsis of this substance consult the work cited above. (Trans.)

Nevertheless, granites have continued very generally to be admitted in many places in the Alpine chain, especially in the enormous mass of which Mont Blanc is the centre and most elevated point. The rock which is there most abundant, has been, and is still very generally called granite.

Geologists had observed, it is true, that this granitic rock of Mont Blanc offered many characters, which caused it sensibly to differ from the granites observed in other chains. It had been remarked that it contained talc instead of mica, that it was very frequently disposed to a schistose texture, and that it sometimes possessed it.

Thus many naturalists have thought it right to describe this rock separately, but always as a variety of granite. M. Jurine alone has proceeded further, and has given it another name, that of *protogyne*, on account of the differences which distinguish it from the true granites. M. Brongniart has mentioned it under this name in his classification of rocks.

The greater number of geologists have not doubtless judged this distinction and denomination necessary, since they have not adopted it. I had long since remarked these anomalies of the granitic rock of Mont Blanc, and they struck me much more when I began to search in the Alps for the characters of the different formations, such as they have been very generally recognized in many countries far distant from each other.*

Besides the presence of chlorite talc instead of mica, and the schistose disposition of this rock, I observed that the quartz was disseminated in it in a peculiar manner, and that it was often wanting.

* Let me be permitted to observe, en passant, that it is rather extraordinary, that we have not as yet a satisfactory geological description of the different primitive formations of the Alps, although it has so often been visited by able geologists. This no doubt arises from the great elevation of this chain, its immense glaciers, and its escarpments, which render geological determinations much more difficult than in low chains.

Another circumstance surprised me much; this was to see, among the debris proceeding from the mass itself of Mont Blanc, the granitic rock associated with many others often extremely schistose, and the greater part talcose; even serpentine is met with, as also masses of actynolite, a substance almost solely belonging to this kind of rock. The true granites of other countries are, on the contrary, nearly without mixture, without any other subordinate rock, and they never, or nearly never have, a tendency to the schistose texture. Lastly, the granite which forms such considerable masses in other chains, appears to me very scarce, and greatly divided in the Alps. There are no doubt chains of mountains without granite; but from analogy, it is extraordinary that true granite should shew itself in such an elevated chain without occupying a greater space, or at least without appearing more frequently.

All these irregularities (many times verified), joined with other circumstances, inclined me to suspect, as I have announced in another memoir (Journal des Mines, No. 137), that the granites and other primitive rocks of the Alps, at least from Mont Cenis to St. Gothard, appeared less ancient than the other primitive rocks; a conjecture which has been adopted by M. von Buch.

Without having the intention of at present attacking in a more positive manner the primordiality of the rocks of the Alps in general, I shall venture to undertake not only to confirm by new considerations the doubts already raised with respect to the granitic rock of Mont Blanc, but to shew that it is not a true granite, either mineralogically, nor geologically, and that in these two respects it appears to unite itself, by different gradations, to a talcose rock widely spread in the Alps.

If I am not able entirely to convince naturalists of this approximation, I hope at least that they will acknowledge that it very well explains the irregularities that the granitic rock of Mont Blanc presents, compared with the true granites of other countries. The talcose rock which I have observed to possess relations to the granite of Mont Blanc, might be named *schistose felspathic talc* or *chlorite*, or, in order not to depart too much from the names under which many of its varieties have been noticed, *talcose (or chlorite) felspathic schist*. The following is its geological association.

It is known how abundant talc and talcose rocks are in the Alps; the mica slates which are there also frequently found, and in the same associations, afford continual passages of mica into talc, from mica slate to talcose slate, and even into the green schist named chlorite schist (chlorite schistoide). It may even be said that talc, and the talcose rocks in general, are much more frequently met with than mica and the micaceous rocks. The micaceous limestones are much more rare than the talcose limestones; and this predominance of talc is above all greater on the Italian side, although it is also observed on the other side of the chain.

Almost all the micaceous or talcose schists are green, and of a green analogous to that of chlorite. Distinct plates of mica or talc are rarely seen in it; it presents more or less shining surfaces, having frequently a slightly fibrous texture.

Among these rocks, to which one would be inclined to give solely the generic name of *talcose schists*, since the greater part are really talcose, and the passage of talc into mica is so frequently seen in the same block; there are some mixed with crystals of felspar, which are the compound rocks, that I name, as I have already said, *talcose felspathic schists*.

They are very frequent in the Alps; I have observed them in Savoy, the Valais, the valley of Aosta, and in numerous places; they are most abundant in the environs of Mont Cervin; and I have since found them in rocks from Corsica. It is impossible that these talcose felspathic schists should have escaped the attention of the numerous geologists who have visited these countries; yet it does not appear that they are often found in collections; neither have they been separately noticed in geological descriptions of different formations, nor in the classifications of rocks; they have no doubt been partly comprised in the gneiss, partly in the mica slates, and partly in the chlorite slates.

M. Brongniart has described one variety under the name of porphyritic gneiss (gneiss porphyroide).

I conceive however that this rock deserves to be considered alone, at least geologically, because it is so frequently met with, and on account of the uniformity observable in its principal characters.

I shall endeavour to notice them, and shew the passages uniting the different varieties.

The principal and prominent base of this rock is a schistose talc, commonly of a green colour, between leek green and meadow green, analogous to that of chlorite, as I have remarked for the talcose schists in general.

The texture is almost always a little fibrous, and the fracture schistose; but the laminæ are scarcely ever so thin as in the true mica slates.

The felspar is uniformly disseminated in it in crystals of most commonly, one millimetre [$\frac{1}{25}$ of an in.] in length; sometimes of two or three centimetres [about $\frac{1}{2}$ in.] in the varieties resembling gneiss (rock of Cervin in the Tarentaise); sometimes very small and scarcely discernible; small white points are solely seen at the surface or rather at the edges of the rock, and the felspar is only recognized in it by the passages that are met with in the same block into other varieties where the crystals are visible.

Quartz very rarely exists in it, and most frequently appears to be entirely wanting; when it is visible, it is very irregularly disseminated in small grains grouped together.

Hornblende does not shew itself in it; at least neither prisms nor laminated needles of that substance are met with; but it is very certainly sometimes intimately mixed in it. Rocks decidedly of hornblende are seen associated in the same mass with the talcose felspathic schist, and insensible passages may be traced between these two rocks.

These and many other passages give reason for conjecturing that it is sometimes to an intimate mixture of hornblende, sometimes to an intimate mixture of quartz, and perhaps also of felspar itself, that the differences of hardness and toughness ought to be attributed, which are observable in the different varieties of this rock.

There are some that are very difficult to break, and at the same time rather hard: these are near the hornblende rocks; others are easily broken, and are very hard, they contain grains and veins of quartz; others, lastly, are tough, as they receive the impression of the hammer without breaking, and are at the same time very soft, so as to be scratched and cut by the knife, like talcose rocks, and especially like chlorite schist, from which in fact these last varieties only differ in the presence of felspar.

From this sketch of the characters of these talcose felspathic schists, it would appear that rocks very different from each other were united under this name; when they are observed in place, we are invincibly led to acknowledge this affinity. When considered in collections, varieties are no doubt remarked apparently very distinct, and which it might even be useful to describe separately, though composed of the same minerals; but these mineralogical differences lose the greater part of their importance when they are not joined with those which are geological, especially when we see, and often in the same mass, very frequent insensible passages of one of these rocks into the other.

I have already noticed the places where I have observed these talcose felspathic schists; if the high valleys of the Alps are passed over, from Mont Rosa to St. Bernard and Mont Blanc, even in part as far as Mont Cenis, and no doubt beyond these limits, a great predominance of the talcose rocks cited above will be met with; and in the places where they are the best characterized and most abundant, will be found serpentine, either pure or mixed with limestone, fine grained hornblende rocks, more or less crystalline limestones, chlorite schists often mixed with oxidulated iron, lastly, the rocks I have mentioned under the name of talcose schists, and in the midst of these two rocks different varieties of the talcose felspathic schists I have described, which sometimes form distinct subordinate beds, and are sometimes united to them by insensible gradations.

Let us now return to the granitic rocks of Mont Blanc.

It has, like our talcose felspathic schists, felspar and talc for its principal constituent parts. The talc in it is most frequently of a dark green, and of the same green colour common to chlorite; it forms small veins which have always a slightly fibrous texture. The rock constantly has a greater or less tendency to the schistose texture; it sometimes even becomes decidedly laminated, and what is very remarkable, the specimens collected at the summit of Mont Blanc afford this character. There is a difference only in an inverse relation between the constituent parts. The felspar predominates in the granitic rocks; the talc in the schistose. But this difference in the proportions, which to the mineralogist is very great, is only essential to the geologist (as I have above stated) when it is joined to differences of position, formation, &c. as between greenstone and syenite.*

Quartz is nevertheless found also in the granitic rock of Mont Blanc; but I would recall to mind that it is sometimes met with in the talcose felspathic schists. It is in truth rare in it; but it is any thing but constant in the granitic rock of Mont Blanc, and it occurs rather in knots or small nodules, scattered or irregularly grouped, than in crystalline grains spread uniformly through the mass, as is seen in the true granites. There are even varieties in which it is altogether wanting, and they are rather numerous.

The two mineralogical differences that I have noticed cannot therefore have any influence on the determination of the geological notions that ought to be entertained of the

* I have cited this example because it is most known. It may nevertheless be with reason objected, that these two rocks are sometimes found united in the same formation; but it is not the less true, that each gives peculiar characters to the formations where it predominates.

Note by the Translator.—The syenite and greenstone of the trap ridge extending from Bolton Beacon to Benton Castle, in Pembrokeshire, form parts of the same mass, as also in the trap district of Gouldtorp Road, in St. Bride's Bay, (Pembrokeshire); the trap in both instances has apparently greatly disturbed the coal measures and carboniferous limestone. For an excellent account of the syenites, greenstones and other overlying rocks, see Dr. MacCulloch's classification of rocks.

granitic rocks of Mont Blanc, whilst on the contrary, the relations already established between this rock and the talcose felspathic schists, make it already presumable that they are two products of the same deposit, in one of which crystallization was more developed than in the other.

If we now add the geological indications I have given to these mineralogical characters, and if we recollect that the rocks of Mont Blanc which have furnished us with these granitic rocks, also afford many rocks decidedly schistose, nearly all mixed with talc, and even completely talcose, since serpentines are met with in it very analogous to other rocks of this kind existing in great masses in the Alps; lastly, hornblende rocks, actynolite, &c. we cannot help recognizing a striking analogy between this association and that which I have shewn to be common to the true talcose rocks, whilst on the contrary this union of rocks has not yet been observed in the true granites.

Lastly, to complete these resemblances, our talcose felspathic schist exists even in the midst of the granitic rocks of Mont Blanc; I have observed it in many places, and particularly near the glacier of Talefre; this talcose schist altogether enters into what is called chlorite schist; it possesses all its characters; it is even mixed with oxidulated iron, but it also contains perfectly formed crystals of felspar.*

This last example appears to me to place the identity of formation as much beyond a doubt as the identity of composition.

We are led then to admit that *the granitic rocks of Mont Blanc can no longer be regarded as granites*, not only according to the mineralogical acceptation at present given to that denomination, but also according to the geological acceptation; they are rather extreme varieties of the talcose felspathic schists I have described, rocks forming a part of the talcose rocks of the Alps; the granitic mass of Mont

* Fragments of rocks are often met with on the sides of Mont Blanc, the half of which is chlorite schist, and the other a granitic rock. There exists in the cabinet of M. de Drée a very beautiful specimen which belonged to Dolomieu.

Blanc, considered geologically, it would appear then must be referred to the talcose formation of the Alps.

I mention these rocks under the name of talcose rocks, and not under that of serpentine, because I consider that they cannot, in many respects, be identified with the serpentine rocks observed elsewhere.

What appeared to me certain is, that they possess very great differences which separate them from granites, and from the relations they bear to serpentine rocks, one is inclined to presume that they are not the most ancient of the primitive class. But until their junction with rocks that are essentially different, has been observed in a positive manner, their epoch of priority cannot be definitively assigned them.

No doubt we might be compelled to form groupes in the midst of these rocks, and to consider separately some members of this formation; but we have not as yet sufficient observations to establish these subdivisions.

I have only spoken of the rocks of Mont Blanc, but I might extend my conclusions to many other granitic rocks of the high Alps, which almost always bear a great analogy to those of Mont Blanc, according to all observers; those I saw at St. Bernard are altogether referable to them; it is the same with many other mountains which follow this last, ascending nearly to Mont Rosa.

This resemblance of numerous granitic rocks of the Alps to a talcose rock, and of the formations containing them to a talcose formation, is founded on a long examination, and continued comparison of the rocks of the Alps, less in collections, than in nature; and it very well accords with the conjectures that I have above brought forward as to the relative antiquity of the greater part of the primitive rocks of the Alps. In fact, if I have before confined myself to suspecting the validity of the titles on which were founded the perogative of antiquity which was accorded to Mont Blanc and the other summits of the Alps, it will be found that I at present destroy that perogative, at least according to received geological ideas, by taking their granite away from them, which being, in the greater number of chains, the base of all the other rocks, formed until now the most plausible character of their primordial formation.

Let us guard ourselves however from pronouncing in too decided a manner on the absence of true granite in the Alps, even in that part of the chain between the Mont Cenis and St. Gothard, the only one that I have visited.

I have observed true formations of granite in the Alps, and their existence appears to me to give a new presumption against those of Mont Blanc and other summits of the centre. In fact, it must not be looked for in these high crests; at least all the granitic rocks that I have there met with approached more or less to those of Mont Blanc (with only some exceptions on which we cannot yet decide); it is in the low mountains which form as it were the advanced posts of the Alps on the side of Piedmont, from Yvrée and even from Turin to the Lago Maggiore. Among the granitic rocks I have observed, there is not one that is analogous to those of Mont Blanc; there are many whose true place I shall not venture to assign; but between Biella and Crevacore, near la Sesia, I have met with a true granite formation having all the characters seen in that of the Limousin, Forez, and other chains. The rocks are never schistose there; the mica is well defined, and does not at all partake of the characters of talc; the quartz is uniformly disseminated; the felspar is often earthy; and the union of these elements constitutes granites, often soft and friable like those of the Limousin. For many leagues I only found half decomposed granites. Lastly, kaolin is there met with, which I do not believe has ever been noticed in any part of the central mountains of the chain, and which appears very generally to belong to granite formations and others that approach them.

I shall add that the form even of these mountains is precisely that admitted as most common in the granite formations; i. e. few escarpments, summits rounded, and as it were in the form of paps; valleys extremely contorted, &c.

There are also, in the neighbourhood, solid granites. The famous rock of Baveno, which has afforded the beautiful crystals of felspar that are so well known, appeared to me to belong to the same granite formation. It is known to be worked much on the great scale, principally for buildings, but also to be cut and polished. There exists yet another working of this kind near Domo d'Ossola, and another near Turin; no doubt the Lago Maggiore, near which the two first quarries are situated, which allows the transport of their products throughout the whole of Lombardy, and the neighbourhood of a capital for the other, facilitates this kind of commerce; but it is rather extraordinary that they should be, at least according to the information I could collect, the only granite quarries in the Alps from Mont Cenis to St. Gothard; and this fact alone gives some reason for presuming that this kind of rock, and consequently the true granite formation, is at least very rare there.

This existence of true granite on one of the sides of the Alps, is analogous to that which has been observed in the Pyrennees, where it is known that the granite most commonly only shews itself at some distance from the centre of the chain.

This character however, taken separately, could not lead to any opinion on the geological nature of the granitic rocks of Mont Blanc; it only acquires importance when it is joined to the direct observations that I have made known. I shall add yet another to it, which is not by itself more decisive, but which appears to me equally to add some weight to my first proofs.

In the granite formation, and especially in those of gneiss and mica slate of other chains, ores of metals are very often found, and generally, more frequently in veins or lodes than in beds or masses.

In the Alps on the contrary, from Mont Cenis to St. Gothard, the ores of metals are scarce, in all the mines that I have seen, and in all the situations I have observed, the ores of lead (at Pesey, Macot, la Thuile, and Cormayeur), those of copper of Olomon, St. Marcel, Servoz, and all the mines of oxidulated iron, are in beds or masses. I am acquainted with but two examples of metalliferous veins that are well determined, one of auriferous pyrites, and the other of copper pyrites; but the latter is met with precisely in the environs of Baveno, near the granite formation, and in the mica slate resting upon it; and the first in a valley (the Val Anzasca), in the neighbourhood of which the same formation occurs.

Some other unworked veins of lead and copper have been noticed, that I could not visit, and on which I might raise some doubts; but many are still, if not in well characterized granite, at least in rocks which I presume to be very nearly allied to it, and not belonging to the talcose formation.

It is at least very certain that metallic veins are extremely rare in that part of the chain I have noticed, in which the talcose formation predominates; and that the only two examples I was able to discover, are, the one certainly, and the other very probably in a different formation.

No doubt geologists have not yet assembled a sufficient number of facts on the occurrence of the ores of metals, so as to assign their existence in beds or veins to epochs relative to the rocks containing them; yet we cannot help observing a great difference between their mode of occurrence in the Alps and other primitive countries, which might seem, in the first instance, to possess analogies to those I have described.

I am fully sensible that it will be necessary to endeavour to observe the junction of the true granite of the Alps, and the talcose rocks of the high summits I have noticed, and under which I have every reason to believe it dips; but it has as yet been impossible for me positively to determine this superposition, and I invite geologists, who may visit the Italian Alps, to endeavour to verify it.

I have not, after all, any want of this last proof for establishing the little relative antiquity of the granitic rocks of the centre of the chain, as it is principally founded on the mineralogical and geological relations of these supposed granites with the talcose felspathic schists, and generally with all the talcose formation so abundant in the Alps.

Perhaps some would raise an objection drawn from the rarity of the granitic rocks in the talcose formation to which they have been referred, and especially from the position of that of Mont Blanc in the midst of the enormous mass of which it is as it were the centre, a position analogous to that very generally given to true granites in other chains.

It may in the first place be answered that this last character ought never alone to serve in establishing a conclusion on the anteriority of a rock, and that it can in no way weaken the proofs I have drawn from the mixtures of this rock, and the associations and gradations that unite it with others; but moreover, what is there extraordinary in meeting with the most crystalline rocks of a formation towards the centre and most elevated parts of the masses that it constitutes? it appears, on the contrary, that reasoning leads one to presume that it ought generally to be so, since these more crystalline masses would better resist all the causes of destruction than others; following up this idea, which is but natural, one is led to conjecture that Mont Blanc remains now the highest and most central eminence of the formation of which it is a part, only because it has been, at its formation, the most crystalline part of it, and consequently the most solid.

Summing up all the geological facts which I have endeavoured to prove in the course of this memoir, it appears,—

1st. That the granitic rocks of Mont Blanc, and others resembling them of the high summits of the Alps (from the Mont Cenis to St. Gothard), are not granites, and that consequently there does not appear to be any true granite in these high crests;

2d. That these granitic rocks are but extreme varieties (more crystalline and more abounding in felspar) of a talcose felspathic rock much more abundant in the Alps, and with which it is found united;

3d. That this talcose rock, also associated with other talcose rocks, constitutes a peculiar formation, predominant in a great part of the Alps;

4th. That ores of metals almost always occur as beds in this formation;

5th. That a true granitic formation exists in the Alps on the southern edge of the chain, which, from analogies founded on all the facts now received in geology, contributes, with all the preceding characters, to establish *the little relative antiquity of the supposed granites of Mont Blanc and the high Alps, as well as that of the talcose rocks of which it forms a part.*

Memoir on the Geology of the Environs of Lons-le-Saunier. By M. Charbaut, Engineer of the Mines.

Read to the Society of Emulation of the Jura, December I, 1818.

(Annales des Mines for 1819.)

The Jura mountains afford the finest field possible for geological observations; yet, it must be confessed, that not only is their structure unknown, but, what is infinitely more vexatious, that prejudices and errors, sanctioned by naturalists otherwise justly celebrated, mislead the geologist who takes them for guides, cause him to commit new errors, or discourage him by their frequent opposition to the facts which offer themselves to his contemplation.

The darkness that yet covers the geological history of such an interesting country, is principally owing to the manner in which, until now, this science has been studied; but it may be hoped that it will be dissipated, since one of the best works that does honour to French naturalists, has given to geology a new direction and range.

When the texture, the facies of rocks, and their chemical composition were the only distinctive characters of formations, the primitive rocks offered infinitely more facility for study than the secondary series; I conceive I am not the only person who has remarked that if we had excellent precepts for the former, geologists had taught us nothing satisfactory concerning the latter.

I imagine the principal cause of this difficulty was owing to the perfect resemblance that frequently exists in the secondary class, between rocks situated at great *geological* distances from each other, and consequently to the confusion resulting from it, of many formations absolutely different.

Since the consideration of fossils in the distinctive characters of rocks has led Messrs. Cuvier and Brongniart to the excellent results contained in their memoir on the mineral geography of the environs of Paris, many geologists have followed this new route; and there is no doubt that from their steps being more certain, they will much sooner be led to the end proposed in geology, which is to understand the relative position of all the mineral masses of the surface of the globe that man can reach.

The Jura mountains, excessively rich in very various fossils, is, perhaps, of all geological sites, the most proper to be examined in this point of view; the most undeniable traces of dreadful convulsions, afford besides, at every step, the image of disorder and chaos; the fractures, the upsetting of beds, the subsidence of entire mountains, cause such surprise, that, notwithstanding the best formed resolution of only verifying facts, the observer is naturally led to ascend to the first causes of all he sees, and cannot avoid deep and prolonged meditation.

M. Brongniart had visited in 1817 a considerable portion of the Jura chain, in a geological point of view, and will probably publish the result of his observations; but as he has not visited the environs of Lons-le-Saunier, I shall endeavour to make known the geological structure of that country.

In order to render this memoir more easily understood, the following is the order I adopt: I commence by distinguishing the great formations; I develope the composition of each of them, beginning with the lowest masses, geologically speaking, and following them in the order of height to those that crown the country, with which I occupy myself; I shew their geographical disposition; and although my end is not to give the mineral geography, I yet notice the nature of the ground in different localities, in order that the facts I advance may be easily verified; I terminate this memoir by noticing some errors until now adopted respecting the composition of the Jura chain. The word formation having received different significations, I conceive it necessary to state that I apply this denomination to all systems of mineral masses, whatever their nature or extent may be, the respective disposition of which proves that they have been formed by an uninterrupted succession of the same causes.

Numerous observations, which many years residence at Lons le Saunier have allowed me often to verify, have induced me to admit two distinct formations.

The first comprises a very considerable thickness of variously coloured marls, which contains many masses of gypsum separated by beds of marl; the whole of these mineral masses is stratified, and covered by a second mass resting regularly on the first, and composed of uninterrupted beds of gryphite limestone.*

The second formation comprises a great height of shelly marls constantly of a slate blue colour passing into ash grey, disposed in parallel beds, which contain some subordinate beds of a shelly argillaceous limes one, often bituminous like the marl near them;† these marls are covered by an enormous mass of limestone generally oolitic, very often passing by insensible degrees into granular and into compact limestone; it is disposed in strata parallel to each other, and conformable to the marl beds in which it rests. The limestone is ferruginous, siliceous, argillaceous, and with or without nodules of flint; occasionally it contains fossil organised bodies.‡

* The first division of this formation will hereafter be seen to be part of the new red or saliferous sandstone formation, and the second division is the lias of our English series. (Trans.)

† These marls are the lias marls. (Trans.)

‡ This mass of limestone apparently consists of the lower oolite division, or the great and inferior oolites. (Trans.)

The second formation does not stop at this limestone, I consider it as being composed of many other successive stages, analogous to the first; but as these do not exist in the environs of Lons le Saunier, I shall only extend my description to the top of the first stage.

Being desirous that my work should accord with that of M. Brongniart, I shall only add that the second stage begins to shew itself at Salins. It is composed, like the first, of shelly and bituminous marls, on which rests a new and very considerable mass of limestone beds.*

The second stage forms the summit of the platform, in which, on the S.E. of Salins, is hollowed out a valley in the form of a gulf; it is seen on the left of this valley resting upon the first stage; the latter disappears on the right, the former sinks in a curious manner at the cascade of Goaille; it afterwards rises, and the eye can easily follow it to above Salins; there remains, near Fort Belin, but an inconsiderable portion, resting against the top of the first stage.

I shall describe the first under the name of gryphite limestone formation, because that pelagian shell, which I have not found in any other position, appears to me characteristic; I shall call the second oolite limestone formation, because that rock appears to me most characteristic of it.

I should observe that the field of my observations being but of small extent, these divisions may not be suitable; generally speaking, I offer them only as provisional in this memoir, waiting until M. Brongniart has characterised these rocks in a definitive manner.

---

* This next stage is probably composed of the Oxford clay and the beds between it and Kimmeridge clay, such as Coral rag, &c. (Trans.)

GRYPHITE LIMESTONE FORMATION.

Rapid examination of the beds it contains.

The first visible beds of this formation are earthy and imperfectly schistose marls, of a deep grey colour, almost black, containing subordinate beds of red gypsum, the fracture of which presents curved bundles of small crystalline plates; the gypsum also forms contorted veins which follow the undulation of the laminæ of marl, and the strings that cut them, and lastly occurs much mixed.

After an unknown but considerable height of these gypseous marls, there exists a thickness of from six to eight metres [26 feet] of compact whitish argillaceous limestone, formed of thin and perfectly even strata.

The gypseous marls again follow, then very marly gypsum, less red than the preceding, the beds of which are separated by beds of marl, subordinate in their turn to the gypseous matter.

Continuing to rise, many masses of gypsum are met with separated by pure marls or gypseous marls; the gypsum, with some few exceptions, becomes more pure; its colour becomes paler, and passes by successive shades into clear red, rose colour, grey, and white; the thickness of these beds increases, whilst the beds of marl that separate them become thinner; the crystalline faces diminish in size, and end by disappearing entirely.

The last mass, furnishing the finest plaster, contains beds of three metres [nearly 10 feet], and even, in some places, five metres [16½ feet] thick; no trace of gypsum is again found to the summit of the last formation of the Jura mountains.

I know not if, as yet, any species of organic remains has been discovered in these masses of gypsum.

There exists above the gypsum rocks a great height of marls, characterised by their varied colours; their section, sometimes discovers in the transverse gorges, ribbon like stripes, marked by different colours, the most general and best determined of which are—white, green, violet, red, grey, and blue; I shall describe them under the name of variegated marls (marnes irisées).

These marls, generally compact and granular, schistose only in the grey and blue portions, are slightly aggregated; they contain more solid beds of different kinds.

A bed of whitish limestone analogous to that which I have noticed in the gypseous marls, but coarser, is first seen immediately above the gypsum rocks.*

There exists, at a distance from the gypsum that I have not yet been able precisely to determine, a bed of very poor schistose coal, the ascertained thickness of which, for a great extent, is between twelve and thirty centimetres [5 inches to 1 foot 2 inches.]

At some metres above, a considerable thickness of beds of a dirty white limestone occurs, divided in every direction by spathose veins, which, resisting better than the limestone the destructive action of air and water, often give it appearance of being divided by partitions.

The variegated marls afterwards contain isolated beds of siliceous sandstone with impressions, of a thickness varying from 3 to 32 centimetres [about 1 inch to 1 f. 3 in.] The organic remains enveloped in them at the time of their formation, appear to have been in such an altered state, that it is extremely difficult to determine their nature.

This sandstone contains nodules and veins of pyrites, the decomposition of which often communicates to it a strong ochre red colour.

Higher up beds of limestone occur almost solely composed of the debris of very small shells; the common thickness of

* These marls will be at once recognised as the upper portion of the new red or saliferous sandstone. A good section of variegated marls will be seen under the lias, from Culverhole to Axmouth Points, on the coast of Devon; between Lyme Regis and Seaton, thin seems of gypsum will also be observed among them. (For sections of this coast see Geol. Trans. vol. I. new series, plate viii.) Trans.

these beds is from 15 to 20 centimetres [6 to 8 inches]; the limestone presents itself as two different species; it is grey, compact, with a splintery fracture; or else it is yellow brown, with an unequal and earthy fracture; in the former case, the stony matter of the shells loses itself in the paste enveloping them, so that they can with difficulty be distinguished; in the second case, on the contrary, the rock is a yellow limestone, from which the shell is distinguished by a chesnut brown colour; I could discover on the surfaces that had long been exposed to the air, some perfect shells of the genera mytilus and venus, some striated pectens and buccina. Before the variegated marls are quitted, subordinate beds are found of a compact whitish argillaceous limestone, of a very fine paste; then the solid beds of gryphite limestone are met with, the name of which I have taken for the entire formation.*

* The gryphite characterizing this limestone, and the Jura Chain, as well in the parts I have visited, as in those from which I have received specimens, is the gryphæa arcuata, (Lam. An. sans vert. ed. of 1818. Knorr. P. 2. pl. lx. fig. 2. Gryphæa incurva. Sowerby, pl. cxii. fig. 1 & 2. Parkinson's Org. Rem. pl. xv. fig. 3.) The gryphæa cymbium much resembles it, and perhaps belongs to the same formation; but I have not observed it in the Jura, properly so called, I have found it in an analogous rock on the coast of Normandy, from Havre to Dives. M. von Schlotheim appears to have confounded under the denomination of gryphites cymbium, the arcuata and cymbium; but they are, notwithstanding their resemblance, two distinct species. The question is here only of the gryphæa arcuata, as determined by the good figures of it above cited; it is the only one I have found in the Jura, properly so called. I have recognised it in many other places, such as the environs of Avalon in Burgundy, near Bayeux and Valognes in Normandy, at Cheltenham in England, &c. and always in rocks that resemble each other in their nature, their position, and the other shells accompanying them.

Note by the Translator.—It is almost needless to remark that the gryphæa incurva, as well as ammonites Bucklandi, mentioned in a subsequent note by the author, as found in the gryphite limestone, are characteristic shells of the lias, and that the rock at Cheltenham is lias, and I can also state that those near Bayeux and Valognes, in which this shell is found, are the same.

I should observe that the argillaceous limestone forms very thin veins in the last beds, almost solely composed of small shells, and in the first of gryphite limestone; its colour is light or blueish grey, according to the rock that it touches.

It results from this observation that the gryphite limestone belongs to the same formation as this species of lumachella, and consequently to the same formation as the immense height of gypseous marls that it covers. This important consequence is confirmed by the relative disposition, and the uniform nature of the beds, on all the points of junction of the gryphite limestone and variegated marls.*

The ordinary texture of the gryphite limestone is compact, but sometimes affords small crystalline grains; in both cases it possesses much tenacity, its fracture is irregular, and its colour blueish grey. It takes its name from a marine shell very abundantly contained in it, of which none analogous are now found living in the present seas.

It besides contains ammonites, belemnites, trochi, turbines, nautilites, terebratulæ, pectens, donaces, venus', muscles, turritellæ, oysters, pinnæ, entrochites, baculites, orthoceratites, &c.†

The difficulty of separating these shells from the rock, prevents the determination of numerous species of them, and of many other genera, of which I could only give an idea by a description of characters insufficient to class them.

* The separation of the new red sandstone and lias is in general very decided. (Trans.)

† Among the shells of gryphite limestone I can only determine the following species:

Ammonites Bucklandi, Sow.; it sometimes acquires a considerable size.

I have found it at Montaines near Salins; to the N.E. of Arau, near Avalon, &c.

The pecten equivalvis, Sow.: which I found with the preceding species, is between Girole and Avalon.

Notwithstanding the labours of M. Lamarck on the terebratulæ, and M. Faure Biguet on the belemnites, it is not yet possible to determine with certainty, and consequently with utility, the species of the two genera found in the gryphite limestone.

The rock that appears to have been the last product of the formation, is disposed in uninterrupted strata, the original and very considerable mass of which has been more or less deeply cut away in its upper part.

The first strata often contain, to the height of 3 or 4 metres [nearly 10 to 13 feet], veins or bands of a kind of calcareous sandstone, which diminish in thickness as they rise, and insensibly pass into the true gryphite limestone. It is remarkable that these beds do not contain a single gryphite, wherever this arenaceous limestone occurs, whilst, where it is not found, thousands of gryphites immediately cover the marl.

At some feet from the arenaceous beds, the limestone breaks easily into straight prisms; the cast of the gryphite detaches itself from the shell, and allows an examination of the interior, where the impression of the muscular attachment of the animal to each valve is distinctly recognised, which rarely happens elsewhere.

I shall lastly observe, that these shells are not mixed and spread indistinctly over every part of these beds; the gryphites are principally assembled in innumerable quantities at their junction; the great ammonites are found laid on their bases among the gryphites; the small ammonites are disposed in groups with the belemnites, without effecting a fixed position in their thickness; the others are less abundant, and are placed in the neighbourhood of these groups.

These beds are not equally provided with fossils; some are so full of them, that the shells compose the greater part of their mass; others, on the contrary, do not contain any, or at least very few; the latter are even and regular, the former receive an uneven and tuberculous form from the large shells that occupy their bases.

Geographical disposition of the Rocks of the Gryphite Limestone formation.

It has been seen from what precedes, that the country, about which I am occupied, does not shew at any point the first beds of this formation ; the rock on which they rest, is buried at too great a depth for observation.

The rocks of the gryphite limestone generally form, at the foot of the first platform of the Jura, the bottom of all the valleys, the greater part of the hills of little elevation, that are rounded and cultivated to their summits, and the base of all the slopes and hills. When they are not covered by the rocks of another formation, they are almost always found masked by alluvium, filling the bottom of the valleys, and by vegetable soil. It follows that their study offers many more difficulties than that of the superior rocks, the numerous escarpments of which readily expose to the observer the disposition and composition of the beds.

It is only by assembling numerous partial observations that we can be enabled to describe them completely. I shall not here undertake to trace all those which have led me to the results I have made known ; but I propose to notice the principal, and the localities that afford the most curious geological facts.

In this respect, the basin in which the town of Lons le Saunier is situated, ought first to engage our attention. To the south of the town, and at a gun shot from Montaigu, on the neck of land separating the basin of Lons le Saunier from that of Macornais, beds of gryphite limestone are seen with the direction of N.N.W and S.S.E. (onze heures de la boussole), and an inclination of 59° to the east; descending to Lons le Saunier, and following their direction, these beds first disappear under the vines, they afterwards shew themselves on the road above, with a thickness of from 25 to 30 centimetres [10 in. to 1 ft.] of limestone without gryphites, with arenaceous veins; an outcrop of partition limestone (calcaire cloisonné), is seen at eighty paces towards the west, among the veins which conceal the variegated marls, and is distinguished at a great distance by its colour, which is whiter than that of the neighbouring rocks.

A point of junction is discovered there between the gryphite limestone and variegated marls.

From this point a narrow hill descends towards the town, composed of the same beds, the direction of which suffers accidental variations, and which lose themselves in the meadow behind the barracks.

I shall en passant remark, that this edifice, though very modern, threatens to fall into ruin, from the foundations reposing on the variegated marls, and because a solid frame work has probably not been formed.

If, from the same point that has been mentioned, the view is directed towards Pimont, on the other side of the town, we always discover in the principal direction of the beds, between the tower and the Villeneuve road, a ridge, not very salient, of gryphite beds, which commences from the meadow, rises to the summit of the knoll (butte), and descends on the opposite side to the village of Villeneuve.

By studying this ridge, the same beds are not only recognized with the same inclination and direction as on the side of Montaigu, but all those are discovered, which compose the variegated marls from the gryphites to beneath the coal bed.

The latter crops out one-third of the way up the hill (à tiers côte) on the side of the road leading to the tower. The administration of the salt springs caused a trial-pit to be driven, twenty years since, which extended about 20 metres [about 65 feet]; its trifling thickness, and the great proportion of earth it contained, caused it to be abandoned.

Beneath the real bed, the remainder of this formation disappears under more modern rocks.

It may be concluded from these observations, that the beds on the slope of Montaigu and the knoll of Pimont, which has just been mentioned, belong to the same gryphite limestone formation; that these beds, evidently moved from their primitive position, have been broken in some manner, so that a very considerable mass, with a mean direction of N.N.W. & S.S.E. (onze heures) has been turned to the E.; and lastly, that this mass has been again cut transversely at the spot intersected by the town of Lons le Saunier. To avoid entering into long details, I shall content myself by noticing that throughout the extent of this basin, from the foot of the platform to Montmorot, the ground is, in a measure, furrowed in similar longitudinal masses, contiguous and cut transversely like the first.

Portions of these masses that are not broken, have remained at the bottom of the basin, uniting at their base the separated hills which enclose it. The town of Lons le Saunier is built on these portions of masses, consequently on the edges of the upset beds of the gryphite limestone formation.

This is the place to mention the salt spring that rises in the town itself.

In no point of this canton is the saliferous gypsum observable, the muriate of soda of which is dissolved by rain water, and gives birth to this spring; it is however certain that it exists in the interior of the formation. I mean to examine if the knowledge of the nature of the ground may not lead to the determination of the place containing the saline rock. The hills in the environs of the town do not expose any outcrop of gypsum, but every thing shews that the knoll of Pimont contains it; I have noticed an elevated ridge of gryphite limestone on this knoll, covering a considerable thickness of variegated marls; now, in all the gypsum quarries of the Jura, the masses of this rock occur precisely under these same marls; it may therefore be presumed that masses of gypsum exist in this knoll, but that they are masked by the debris of a more recent formation, which covers the lower part of the variegated marls, beginning with the coal bed.

This conjecture acquires additional support from the existence of a small spring of selenitous water, which breaks out precisely at the lower part of the variegated marl beds. I shall here state that this spring supplies the fountain of the abbey, the waters of which are known in the town not to be fit for dressing vegetables or for the purposes of washing, owing to its containing sulphate of lime.

It is then very probable that the knoll of Pimont contains gypsum. On the other hand, the invariable abundance and elevation in the valley of the saline spring, shew that the waters come from an elevated place offering a large surface to the rains.

Now, the knoll of Pimont, which of all the surrounding hills most indicates the presence of gypsum, is still the only one, within a radius of more than three kilometres [12,000 feet] that unites these two conditions; it may then be presumed that it contains also the saline rock traversed by the rain water before it arrives at the spring.

It will not be misplaced to mention here a phenomenon rather frequent in the Jura, which is essentially owing to the formation I am describing; I mean of the funnels which suddenly form, by the sinking of the surface of the ground.

The machines for the extraction of the salt water are placed at the bottom of a vast funnel 16 metres [about 52 ft.] deep, which has originated in a sinking of the ground, at an unknown but no very distant period.

In 1792, a considerable sinking took place in the Rue des Dames; the inhabitants of the town were seized with the greatest fear, when they saw a house gradually sink and disappear in a gulf which was immediately filled with water.

Witnesses worthy of credit state, that the waters of the saline spring suddenly became low, and were greatly disturbed during this event.

Lastly, there recently occurred between the two former, a third, but much less considerable sinking; although I was not at that time present, I nevertheless learned with certainty, that the spring was again disturbed by this cause.

The explanation of the fact is very simple; it can be conceived, without having recourse to any hypothesis, that the systems of enormous beds, when upset into the positions in which we now see them, could not coincide so exactly with the inferior rocks, as not to leave some empty spaces between them.

The water, while traversing these subterranean passages, must, by its motion and the pressure resulting from the more or less considerable elevation of the springs that feed it, continually wear away their sides, and in the end cause sinkings that sometimes extend to the surface.

Lastly, when it is observed that ten pumps constantly employed to obtain the waters of the Lons le Saunier spring greatly augment the rapidity of their subterranean course, it will not be surprising that this event should occur more frequently at this point than any other.

From observations made during these sinkings, on the saline spring, it cannot be doubted that at least a part of its waters does not reach them. Now, it is precisely the direction in which the gypsum beds of the knoll of Pimont ought to occur: this fact renders the existence of saliferous gypsum infinitely more probable in the heart of this hill.

Two other saline springs, known by the names of Cornoy Pits, and the Saloir Pool Pits, form, with the above, the objects of the Montmorot salt spring works.

These are situated at 3 kilometres [about $2\frac{1}{3}$ miles] to the west of the former, on the hill opposite the knoll of Pimont, and separated from each other by a longitudinal hill of the gryphite limestone formation.

Two trial pits, cut in the variegated marls, have shewn the existence of gypsum on this hill.

The springs of Salins, worked at the two salt-works of Salins and Ure, are situated in the same formation; on the right side of this valley the finest escarpments of the gryphite limestone formation are seen, and the best gypsum quarries of the Jura mountains worked.

While terminating my observations on the position of the salt springs of this department, I shall mention a specimen of roseate saliferous gypsum, brought from the commune of Toulouse, situated at two myriametres [more than 15 miles] from Lons le Saunier, which was sent me by M. Bichet, ex-secretary of the prefecture. My researches to discover its position have as yet proved unsuccessful; but numerous indications give me hopes of finding it in the end.

Lastly, two small salt springs have been mentioned to me as occurring in the communes of St. Lauthain and Tourmont; a quarry of gypseous alabaster was formerly worked in the former.

These indications are perfectly in accordance with the nature of the rock which serves as a base to the ground of the three communes; and it is remarkable, that their position, with regard to the springs of Lons le Saunier, is at a small distance from the general direction of the beds.

In the upper part of the valley of Lons le Saunier, the gryphite formation totally disappears beneath the oolite formation; but it begins to rise below the village of Conliege, where it forms small hillocks, which have a direction from S.E. to N.W., and it is easy to distinguish the neighbouring rock solely from their aspect.

I here state that the gryphite limestone formation is beneath that of the oolite limestone. I should premise that this important allegation is not the result of doubtful observations, but that I have already seen, by the aid of many sections, at Montaigu, Montmorot, &c. the rocks of the second formation cover those of the first; the suite of this memoir will moreover shew that not a step can be made in the environs of Lons le Saunier without acquiring proofs of this fact.

At the foot of the village of Perrigny, a bed is seen in which the gryphites are transformed into pearl spar; their white colour contrasts with the base of the rock, coloured dark red by oxide of iron: it is placed five metres [about 17 feet] above the variegated marls.

This peculiarity is only interesting because a similar limestone is found at great distances from each other, belonging no doubt to the same bed. At the end of the meadow, and from this point to Lons le Saunier, the gryphite beds have still a S.E. and N.W. direction, that is, the same as this portion of the valley; but they again acquire, at a short distance, their general direction, which varies little from a N. and S. line.

The most elevated hill met with on the left of the road from Lons le Saunier to Pannessieres, is the only one, where I have as yet been able to recognize in place the beds of siliceous sandstone with impressions.

The coal bed of the gryphite limestone formation appears on the summit of the Savagnu hill, on the south of the village.

The knoll of Montmorot is based on a small isolated portion of the longitudinal hill separating the two neighbouring salt springs, cut on the south by the principal valley, and on the north by a small transverse gorge.

This isolated knoll, around the foot of which the gryphite limestone formation is recognised, is crowned by beds of the oolite formation ; this point offers an undeniable proof of the proposition I have advanced.

On the north of the knoll, a fine breaking away (arrachement) of variegated marls is discovered, the beds of which run nearly N.N.W. and S.S.E. (11 heures), and inclined towards the east at 50°; it is in these marls, at 200 metres [656 feet] to the N. of the gorge, that two pits have been driven to the gypsum; the dark red colour of that which I collected among the debris, leads me to think that it belongs to the inferior masses of the formation.

The two gypsum quarries of Courbouzon and St. Laurent are situated on the prolongation of the direction of these beds; every thing leads me to believe that they are hollowed out of the same longitudinal hill (coteau), which disappears at Mont Orient, beneath the oolite limestone formation.

The gryphite limestone rocks of the valley of Macornais are but the prolongation of the hills in the basin of Lons le Saunier, which are covered in the interval by patches of the oolite limestone formation.

The coal bed which I have already noticed at two points in the Lons le Saunier basin, occurs again in the valley of Baume a Voiteur, where an excellent gypsum quarry is worked; it belongs to the same formation.

This small bed of coal did not appear to me worth working at any point; but it is very remarkable on account of its constant inferior position to the gypsum masses, to which it may serve as a guide, and by its great length; it has been recognized as far as the department of the Doubs.

I shall not further extend my observations on the gryphite limestone formation, in order to arrive more speedily to the following

---

OOLITE LIMESTONE FORMATION.

Examination of the beds composing it.

The marls of this formation differ from the preceding by their constant grey or slate blue colour, by their much more slaty texture, often by their bituminous nature, and by the fossil shells they contain.

The lowest are compact, fragile, and of a slight grey tint; they contain a very great number of small ammonites, belemnites, entrochi, some small pectens, and terebratulæ; the shells are found changed into pyrites. These first marls contain nests of cubic pyrites, and oxide of iron, arising from the decomposition of the sulphur.

Higher up, fewer ammonites are found; the marl becomes earthy, and contains thick beds of marly limestone, in which small bacculites truncated in the manner of entrochi are seen.

After a great number of beds with few shells, one is met with of prodigious richness as to species; belemnites, pectens, and anomiæ, grouped in families, form alternating strata, in which the cementing rock is scarcely discernable; among other less abundant shells that it contains, I may notice ammonites, donaces, nautili, mytili, baculites, entrochi, and cardiæ.

The grouped shells are generally small; but isolated belemnites 15 centimetres [six inches] long are found, and ribbed pectens of the size of the hand, that are easily detached from the rock; it is remarkable that the small bivalve, which I consider may be referred to the genus anomia, is very rare in the Jura every where but in the marls of the oolite formation.

Between the beds, numerous nodules of tenacious marl exist, possessing very remarkable peculiarities; they are solids generally of a cylindrical form, the axis of which, from 8 to 10 millimetres [less than ½ inch] in diameter, is a small tube of iron pyrites, often decomposed, filled with calcareous spar; they resemble in form columns, ballustrades, eggs, fruits, *&c.*; some are found more than a metre in height, their position is nearly vertical; the latter tube is sometimes forked; two are very often joined together, and the form of the solid is always modified in consequence.

No fossil shells are seen in these detached bodies, whilst on one side, the tubercular nodules, without central axes, are filled with them; among the shells are found, venus, planulites, ammonites, belemnites, trochi, turritellæ, entrochi, baculites, and rarely pectens. They are so well preserved that they appear as if just taken from the sea; some are covered by a very thin bed of sulphuret of iron, which gives them a very beautiful bronzed appearance.

Lastly, very large irregularly rounded nodules are met with, of a compact and brittle marl, the interior of which affords cracks as much as five centimetres [two inches] wide, lined and often filled by sulphate of strontian, of a white or light sky blue colour.*

After having observed these nodules, situated on the same marls, we know not which to be most surprised at, whether

* The sulphate of strontian in voluminous crystalline masses, of a very lengthened plate, but extremely brittle, and of a dirty white, more or less blueish, appears to belong to the inferior masses of the oolitic limestone, and commonly occurs either in cavities of the limestone, or in those of the large shells, such as ammonites.

The analysis made by M. Bertier has verified the nature of this substance. I have seen it absolutely in the same position lining the cavities of very large ammonites in the bed of Ergoltz, near Liestal, in the environs of Basle, accompanied by belemnites, and a shell that appears to be the Lima antiquata of Sowerby.

the regular and constant form of the first, the abundance of various shells, as it were exclusively found in the second, or lastly the existence of very pure crystals of strontian in the midst of the latter, the mass of which does not appear to contain any of the constituent parts of that substance.

The explanation of these facts is yet a problem, the solution of which would be most interesting.

Above the very shelly bed, the marls become bituminous, and contain an innumerable quantity of small flat bivalves, of the breadth of a lentil, only distinguishable by their transverse and almost circular striæ.

The action of the air in the first place divided these marls into thin laminæ, resembling the finest slate, and afterwards reduces them to powder.

Besides these microscopic shells, laminæ with impressions of very flat ammonites and planulites (planulites), are met with, as also some small entrochi or baculites.

Tenaceous beds from 12 to 24 centimetres [about 5 to 10 in.] thick, occur in the midst of these bituminous marls, containing small veins of black and brown bitumen, divided in a rectangular manner by very fine white spathose partitions;* the rock itself is impregnated by a great quantity of bitumen; impressions are there seen of wide planulites (planulites), anomiæ, a thin and brittle triangular bivalve, and many other genera of shells as difficult to determine as the latter.

I observed a height of more than 20 metres [about 65 ft.] of these bituminous slaty marls, and I have not perceived their upper termination, which proves that they are still more extensive.

Marls without bitumen succeed them, and rise to the first oolite beds; they are imperfectly slaty, and in a great measure earthy; they contain at their upper part beds of shelly and marly limestone, between which are found large ammo-

* A flattened kind of black fossil wood, which splits into rectangular pieces, separated by thin spathose partitions occurs in the lias marls of Lyme Regis, as also detached flat pieces of a similar nature, the vegetable structure of which is not apparent. (Trans.)

nites, planulites, and nautilites, disposed parallel to the beds, forming a kind of nodule easily separated from the marl.

A bed of oolite, composed of small grains of oxide of iron cemented by marly limestone, regularly rests on this enormous mass of marl beds; the grains are yellow and red, their form resembles that of gunpowder; the cement is often ferruginous, its colour is then reddish brown; at other times it is grey, like that of the inferior marls. This oolite has not much tenacity, and the action of the air reduces it to powder, but it is more durable in proportion to the oxide of iron it contains. Its richness is sometimes so great that it is worked as an iron ore.*

The common thickness of this bed is two metres [about 6 feet 6 in.]; it varies to five [about 16 f. 6 in.].

Among the numerous shells it contains, I may notice planulites, belemnites, ammonites, terebratulæ, pectens, trochi, donaces, nautili, and entrochi; the two first are the most abundant.†

Crystals of sulphate of strontian occur in the geodes of calcareous spar existing sometimes in the middle of these shells, especially in the last whorls of the planulites.

Above this ferruginous bed some beds occur still separated by marl strata, forming the passage from grey compact marly limestone into yellowish siliceous limestone; the marl afterwards disappears entirely; the passage is even sudden where the ferruginous bed is thick. Here commences that long suite of oolitic granular or compact limestone, composing the escarpment of the first platform of the Jura.

From the idea attached in mineralogy to the denomination of granular limestone, it may appear surprising to see this rock among the secondary formations; it is nevertheless certain that limestone beds, composed of small lamellar

* This is evidently the inferior oolite. (Trans.)

† For a detailed list of fossil shells contained in the English inferior oolite, consult Messrs. Conybeare and Phillips' Outlines of the Geology of England and Wales, p. 239, &c. (Trans.)

grains are found in the Jura, which are extremely difficult to be distinguished at first sight from primitive granular limestone; but in the fracture traces of compact portions occur mixed with the crystalline parts.

The compact portions commonly form the base of the rock, and the lamellar grains are only subordinate.

In the oolitic limestone, the size of the grains varies from one bed to the other, from that of a rape-seed to a nut; but they are nearly all of the same size in the same bed.

I should observe, that, from the form of its grains this rock was more nearly allied to pisolite than oolite, but as the grains generally are very small, the first denomination would be improper in the greater number of cases.

However small the grains may be, concentric layers are almost always seen in them round a nucleus, often crystalline; this nucleus is sometimes the fragment of a baculite, entrochus, or some other marine shell: now these organic bodies could not have been lapidified and enveloped by stony layers before the formation of the rock; on the other hand, whatever may be the nature of the other nuclei, the grains enveloping them have probably been formed in the same manner as the others. I conceive then I can with some certainty advance, that in the Jura oolites, the grains did not exist before the formation of the paste, and that this rock has not been formed in the manner of conglomerates.

The oolitic and granular limestones alternate, and pass into one another by insensible gradations.

It contains various proportions of silex; numerous veins and irregular nodules of black and white flint (silex) occur in it, which, by long exposure to the air, are often transformed into spongiform quartz.

In numerous beds, and on the surfaces slightly corroded by the atmosphere, an innumerable quantity of debris of very minute marine bodies are seen, such as entrochi, orthoceratites, baculites, vermiculæ, ramified polypi, sponges, &c. These fossils are so much identified with their matrix, that none can be distinguished in fresh fractures of the rock; some large pectens, belemnites, and terebratulæ can only be perceived.

At a considerable height, beds of granular limestone are found full of circular and small smooth pectens, of the size of a lentil; afterwards a grey compact limestone always with flint nodules, containing large imbricated pectens, and a longitudinally striped bivalve, referable to the genera mactra or venus. The valves of this shell, without being separated, appear to have been turned over the plane of their junction, so that the teeth no longer correspond; large bivalves are also seen with deep folds, the genus of which it has as yet been impossible for me to discover.

Among these last beds there is one, the rock of which perfectly resembles the gryphite limestone, and which is surprisingly rich in marine fossils; besides the greater part of the shells that I have already noticed, I have found very fine orbulites in it of the size of an egg, which I have not elsewhere met with; an undetermined fragment of an organized body of the size of an arm, having a bony texture, and in this respect analogous to a much smaller fossil, which I found in the department of the Doubs, and which bears some resemblance to the spine of a sea fish, of a species of sting ray (raie aigle);* fragments of a very large bivalves are seen, the shell of which is more than five millimetres [$\frac{1}{5}$ inch] thick, with traces of many other animals.†

The compact limestone afterwards becomes very siliceous, and contains no shells, for a great height; it is transformed

* May not this be the same kind of bone which I have described as the radius of some fish in the first vol. Geol. Trans. new series, p. 43, 44, and figured pl. 4, different species of which are found in the transition limestone, carboniferous limestone, lias, oolite formation, and chalk? (Trans.)

† Among the shells found in this rock, and of which very few of the species are described, or in a state to be determined, ammonites discus of Sowerby may be noticed, which I have found in the ferruginous oolitic rock of Aisy in Burgundy. The pecten lens, Sow. is also found in Burgundy in the same rock and with the same shells.

into beds of conchoidal flint (silex conchoïdes), it afterwards re-appears in regular and very thin beds.

Above it are seen, first in a compact limestone, afterwards in a granular limestone, very beautiful, smooth, and striated terebratulæ, of the size of a pigeon's egg; they are assembled in groups, and adhere but slightly to the rock.

They are associated in the granular limestone with entrochi and echinites; the entrochi are so abundant, that they form the greater portion of many beds; terebratulæ, not so well preserved as the former, are again mixed in it, with certain undetermined bivalves, of which the interior cast only is found.

Beds of large grained oolite occur at this height; around the grains, the rock forms concentric layers, which leave no doubt of the contemporaneous formation of the grains and their matrix.

This rock is sometimes hard enough to receive a good polish; its colour is grey, slightly blueish, but the edges of the bed and fissures are of a yellowish white colour; this last tint is produced by the alteration of the former, by the in-increased oxidation of the iron it contains, wherever water has been able to penetrate.

This alteration of colour is seen in many quarries of building stone in the Jura; a kind of natural ornament in the buildings results from it, producing a very beautiful effect. It may particularly be observed in the town of Besancon.

To the limestone full of terebratulæ and entrochi, succeed beds, slightly shelly, of white oolitic and granular limestone, furnishing the best building stone in the country.

Quitting these beds, no more granular limestone is found to the second stage of the oolite formation.

Very thick beds of compact limestone alternating with oolites, constitute the top of the first stage.

This compact limestone is white; its fracture, at first undetermined, becomes, as it rises, more and more conchoidal; in the upper beds it is brittle, and its fragments are very sharp.

A zoophite, which I refer to the genera astrea or tubipora, with regular hexagonal tubes, is frequently met with in the compact limestone of an undetermined fracture; some pectens and terebratulæ are also seen in it.

The last beds of brittle limestone do not appear to contain any species of organic remains.*

On the last of these beds rest the marls of the next stage; but I do not propose to describe them here, because they do not occur, as I have already stated, in the environs of Lons le Saunier.

Geographical Distribution of the Rocks of the Oolite Formation.

The rocks whose composition has just been described, form the first step of the immense amphitheatre presented by the Jura chain on the side of France; the road from Lyons to Strasbourg, for a length of four myriameters [about 30 miles] on each side of the town of Lons le Saunier, is traced precisely at its foot in a direction from S.S.W. to N.N.E.

To the south of the town, the edge of this step is deeply furrowed and divided into longitudinal hills covered to their summits with wood. On the north it presents a vast platform cut by some transverse valleys, and bordered by well cultivated knolls and hills; these hills, of small elevation, and situated on the prolongation of the slopes, are almost always crowned by a few beds of yellowish rock of little extent.

Observing the sides of the valleys with care, I recognised throughout their whole extent the succession of beds I have noticed.

The marls are nearly every where masked by the vegetable soil; they are only seen in some broken places, so that

* The beds described by the author between the inferior oolite and the first stage, are most probably analogous to the great oolite, &c. of the English series. (Trans.)

their first inspection might leave doubts on their prolongation beneath the mass of oolitic limestone; I had long observed them before I could decide this question; but I at last remarked in some places, and especially to the E. and S.E. of Conliege, an insensible passage of the marly limestone beds into the oolitic beds; now, the first are subordinate to the marl beds; it is then evident that they all belong to the same formation. I had moreover occasion to make excavations beneath the ferruginous bed, and I always found the marls, which leaves no doubt of the existence of the marl beds beneath the solid beds of oolitic limestone.

The perfect identity of the beds on each side the valley proves that they have been united, and that they originally formed but one mass.

The nature and respective dispositions of these mineral masses being well understood, the manner in which the valleys have been formed may be concluded, without having recourse to any hypothesis.

In fact these mineral masses have been deposited by the waters of the sea: this fact is proved by the marine animals found throughout their whole extent. When the waters abandoned them, the marls must have given way under the weight, now become more considerable, of the mass that covered them; their slight tenacity offering less resistance than that of the upper masses, the waters carried them away and excavated passages in their mass, forming vast galleries.*

If we represent to ourselves the considerable thickness of the marls, in some places 100 metres [328 feet], the following picture presents itself to the imagination:

An immense platform, under which vast subterranean passages exist, which the waters continue to enlarge, sinks as

* The character of these valleys would appear to be analogous to the valleys on the edge of the oolite escarpment near Bath, &c. For an account of the valleys of denudation on the south coast of Devon and Dorset, consult Prof. Buckland's memoir, in the 1st vol. new series, of the Geological Transactions. (Trans.)

the marl is carried away from beneath it; the portions deprived of all support giving way to their enormous weight, descend and break; their debris are washed away by the waters, rolled to a distance, and deep ravines or valleys are left in their places.

The better supported portions resist, but are mined at their feet, their beds projecting over the edge of the ruptures, are upset, are placed in inclined positions, and are curved and contorted according to the form of the rocks on which they rest. From thence the mountains and the various forms taken by their beds.

The caverns occurring on the oolitic mass, are owing to some of these galleries, the roof of which being better supported, has only fallen in part. They are true subterranean valleys.

I have not the vain pretension of creating an universal system. I merely state that I apply my reasoning solely to the rocks in which I have observed the facts on which it is founded.

The valleys of Macornais, Lons le Saunier, Voiteur, &c. form, at their commencement, the furrows in the platform I have mentioned: they are afterwards prolonged across the slopes and hills.

The general aspect of their elevated sides is that of a wall of rocks, which is surmounted by a rocky and almost vertical formation, covered by heaths, and from which descends with an uniform slope a fertile bank, covered with excellent vines.

The line of junction of the rocky wall with the kind of prop commencing at its foot, is precisely at the separation of the two enormous masses of limestone and marl; we are always certain of finding the ferruginous oolitic bed there; this line follows the slight inflexions of the beds, that are generally almost horizontal.

The action of the air and water decomposing the marls that support the limestone beds, would mine the foot of the escarpment until the fallen masses covering the marls with debris should protect them from the atmosphere; since then the talus would be covered with vegetable soil, and would only from time to time be exposed to the falling masses.

If it is observed that the calcareous mass is split in all directions perpendicular to its strata, and that it breaks in right lined prisms, it will be seen why the escarpments are always perpendicular to the beds; and the regular and constant form presented by the sides of the valleys will not be at all surprising.

The upper part of the third valley above mentioned, known by the name of Roche de Baume, appears much more recent than the others; the escarpments offer considerable precipices, from which frequent falls occur; the debris form a perfectly regular talus of 45°, which is not any where covered by vegetable soil; and it is probable that ages may elapse before it can be cultivated.

The form of these valleys present the same sinuosities as a serpentine river; the salient angles of one side corresponding with the re-entering angles of the other. If these partial deviations are disregarded, there are but two principal directions, one from S. to N. the same as that of the longitudinal vallies, and the other from S.E. to N.W.

It is remarkable that when the first direction is changed into the second, there is generally found on the east a small accessory valley, rounded in the form of a gulf.

This fact may be observed in the valley of Macornais, opposite Moyron, in that of Lons le Saunier opposite Ravigny and Conliege, and in that of Voiteur, opposite the Abbey of Baume, and above Nevy.

Does not this shew that the waters of the sea retired from this country in a general direction from S. to N.? that the interior currents formed in the midst of the marls, having met with obstacles in certain points, had in their course rushed from right to left, but not being able to force a passage towards the east, they only formed a kind of circus there, returning upon themselves, so as to throw themselves wholly towards the N.W.?

The entrance of a deep cave is perceived in the small gulf of Revigny, composed of immense subterranean chambers, situated in a north and south direction, and communicating with each other by passages of different dimensions.

The river la Seille, flowing at the bottom of the valley of Voiteur, rushes out of a cave perfectly resembling the last.

If it be observed that the latter is situated precisely on the north of the former, that is to say, in the direction in which the waters of the sea have retired, it will appear very probable that the two caves communicate with each other, and form but one.

A kind of subterranean lake prevents the verification of this conjecture.

It is very probable that the rocks of Baume, mentioned above, form a continuance of the cavern of la Seille, and that they have fallen down much later than the other excavations of the valley of Voiteur.

The calcareous mass of this formation is traversed throughout its height by very numerous clefts; it follows that the rain waters are not retained on the surface, they infiltrate down to the marls, and give rise to many springs always situated at the foot of the rocky escarpments.

This is the cause that there exists neither fountains nor rivulets, throughout the extent of the platform, which compels the inhabitants of the villages to consume only cistern water.

It follows also that the ground is lighter and more fit for the culture of corn.

Lastly, the waters, penetrating into the caverns, gradually wear away their sides, and in the end produce sinkings resembling those I have described in the gryphite limestone formation.

The first stage of the oolite formation is not complete in the vicinity of Lons le Saunier; the platform rises at the commencement of the valley as high as the beds containing entrochi and terebratulæ above Conliege, to the thin and regular compact limestone beds that precede it; it is still lower above Perigny.

It attains at St. Maure, Crançot, and nearly to Mirbel, the height of the oolitic and granular beds, which afford the best building stone.

Beds of compact limestone, with an undetermined and conchoidal fracture occur near the latter village; they produce the most esteemed lime in the country.

The valley of Lons le Saunier, after having been enclosed to the limit of the platform, suddenly opens to the right, near the village of Perigny.

This limit, distant 2 kilometres [about 1½ mile] on the east of the town, passes by the villages of Pannessière, Lavigny, Voiteur, and Frontenay; it afterwards turns a little to the east towards Poligny and Arbois.

The platform is prolonged, on the left of the valley, to Montaigu, forming a rounded projection which separates the two basins of Lons le Saunier and Macornais.

Beyond the limit of the platform, the basin of Lons le Saunier is bordered on the left by four longitudinal hills, having a N. and S. direction, which are united together, and to the projection of Montaigu by passes of little elevation; it opens out to the right, and is bounded by very flat hills.

The oolitic rocks have, in this part, been almost entirely carried away: some scattered patches only remain on the gryphite formation, forming the knolls of Montmoret, Pimont, l'Etoile, the hills of Chilles, le Pin, Montin, Plainoiseau, Arlay, &c. These are, it may be said, the witnesses of the great convulsion that has torn away the platform which covered all this country.*

The gryphite limestone rocks have been laid open throughout the greater part of this extent; I shall observe, that in all situations where the junction of the first marls of the last formation with the gryphite rocks can be observed, they are found disposed in beds parallel to the portions of beds on

* These patches seem to be outliers resembling those of the same rocks in the vicinity of Bath, &c. (Trans.)

which they rest, and divided into small prisms perfectly resembling a pavement.

This observation had, for some time, shaken my opinion respecting the distinction of the two formations; but I have since convinced myself, that this parallelism only belongs to the first beds deposited on the gryphite formation, and that soon after the marls no longer present this relationship.

The most complete proof is obtained of the difference of formation of these two kinds of rocks, by seeing the most ancient beds break through the foot of the talus of shelly marls, between Conliege and Perigny; also breaking through above the bituminous beds of the knoll of Pimont, with an inclination of 50° to the east, whilst the latter dip in a contrary direction; by seeing them rise nearly to the oolitic beds at Montaigu, to the marls of the second stage at Salins, and much higher still in other places.

I shall lastly add, that immediately on the N.E. of Lons le Saunier, perfectly horizontal beds of pyritous shelly marls are found, and that at less than a gun-shot to the north of this point, vertical beds of gryphite limestone are seen rising above the ground, without varying sensibly in situation; which clearly shews that the pyritous shelly marls, and consequently all the oolitic rocks, did not exist at the time of the displacement of the gryphite beds.*

This is the place to describe the singular positions in which the upset beds are found of the enormous masses of oolitic limestone; but not to prolong this memoir to too great a length, I shall not undertake to treat on this subject with all the detail of which it is susceptible; I shall confine myself to citing the following examples:

* These appearances may very easily be produced by faults, known not to be uncommon in lias, the coast section I have given in the Geol. Trans. vol. i, new series, plate 8, shews some that have affected both the lias and new red sandstone beneath it. That considerable faults occur in the neighbourhood of Lons le Saunier, is clearly seen by the author's account of that country, therefore the different beds may easily come in contact with different dips. (Trans.)

The projection of Montaigu with the first longitudinal hill joining it, forms a creek situated on the prolongation of the first direction of the valley of Macornais; the waters, descending from Vernantois, have upset the beds of this slope towards the E., and carried away into the basin of Lons le Saunier, the oolitic marls which covered the projection, leaving as a witness of their existence, some remains of beds at the end of the village on the gryphite rocks; the waters not being able to break the latter, have been compelled to return upon themselves, and to cut, in order to enter into the valley of Macornais, through a considerable mass of beds which were united to the upper part of the hill, and which, at present, form an isolated knoll opposite the village of Moyron.

Beyond Macornais the valley opens considerably to the right; its waters, thrown back by the current, descending from the village of Vaux on the opposite west side of Munsy hill, have upset its beds to the W.S.W.

There is very little inclination on the part of this hill looking upon Lons le Saunier; but it always increases towards the other extremity; in the quarry of Paradis, situated opposite Macornais, beds of entrochi and terebratulæ limestone are seen inclined at more than 45°.

The beds of compact and oolitic limestone covering these last, and which very much resemble the brittle limestone, form on the west a knoll, the stratification of which is much confused, and which is not altogether detached from the hill.

It is very remarkable that this species of promontory, from whence a view is obtained over the whole breadth of the plains of the Saone, rests immediately on horizontal beds of gryphite limestone, an escarpment of which, from six to eight metres [about 20 to 26 ft.] in height, and 100 [328 ft.] in length, crops out among the vines at the foot of the promontory.

This fact proves in the most decided manner that the disturbances of the rocks of the oolitic formation cannot be owing to a cause from the interior, such as an earthquake.

Mont-Orient, situated to the south of Lons le Saunier, behind the village of Courbouzon, presents very curious disturbances of the beds.

This mount intercepts the communication of two long villages, being situated in the line between them, and presents a very elevated escarpment on the side of Courbouzon, of two systems of beds upset to the E. and W. on the bottom of the lower valley, and resting against its sides, the beds of which are inclined towards the centre on the left, and are horizontal on the right.

To account for this singular disturbance, it is sufficient to imagine that the waters that descended from St. Laurent towards Courbouzon, having formed, in this broad part of the valley, two galleries instead of one, separated by a mass, either of gryphite beds, or marls more solid than the rest, in each side of which the sinking took place as in two different vallies, and that beyond, the two united currents entirely carried away the beds, which were thus left unsupported.

These beds, leaning against each other, opposed such resistance to the waters, that they were compelled to open a passage to the west, near the village of Gevingé.

The beds of the right side of this mount occur disturbed under various angles; they are first inclined at about 40°, and afterwards at 75°.

Lastly, these compact and oolitic beds are in a perfectly vertical position; but those on the left have sunk more than the former.

The country house of the late Peer of France, Vernier, is situated on the summit of this mount, at the separation of the two systems of disturbed beds.

At its foot is worked, in the gryphite limestone formation, the gypsum quarry of Courbouzon.

I ought to state, before I terminate this memoir, that the two rock formations of the Jura chain have already been noticed by several geologists; but I should, at the same time remark, that they considered that which contains gryphites, as more modern than that of the *true compact limestone of the Jura, which easily breaks into conchoidal splinters.*

The cause of this error was owing to the respective extent and limits of these formations not being known, organic remains were supposed to be extremely rare in the second, and very abundant in the first, and lastly the beds were considered to be more horizontal in the latter than the former; so that the limestone formation, the beds of which were arched, appeared referable to the transition series, whilst that of the gryphite limestone was regarded, and with reason, as a secondary formation in all the force of the term.

The most important geological fact that I have proved in this memoir, and of which the hollowing out of the valleys and caverns of this country ought to be a necessary consequence, was unknown. I speak of the existence of enormous masses of marl beds, beneath the gryphite limestone, and beneath the different stages of oolitic limestone; in order to explain these great effects, recourse must be had to purely gratuitous hypotheses.

Lastly, these marls, which always appear as if overlapping, either at the bottom of valleys, at the foot of platforms, or on their summit, not having been distinguished, as to their formation, from those found in irregular beds among the clays, sands, and gravelly soils; they were considered as deposited after the formation of the valleys.

It followed that the gypseous masses of the Jura, which are the most ancient of all the mineral masses of that chain, appeared to be of an infinitely more modern formation.

This memoir is very incomplete, and partakes greatly of the haste with which it was written; but the subject is so vast, that it would require many years of observation and study, in order to understand the immense quantity of fossils that exist in the environs of Lons le Saunier, and properly to describe the rocks bed by bed.

The much more limited end, that I propose, will be accomplished, if this memoir may serve to dispel some errors, that were entertained with regard to the geological structure of these mountains, and it may, in the end, render their study more easy.

Fig: 1.

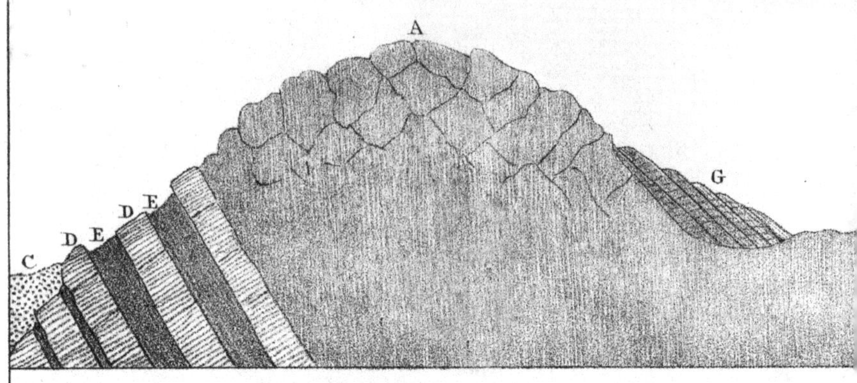

Fig: 3.

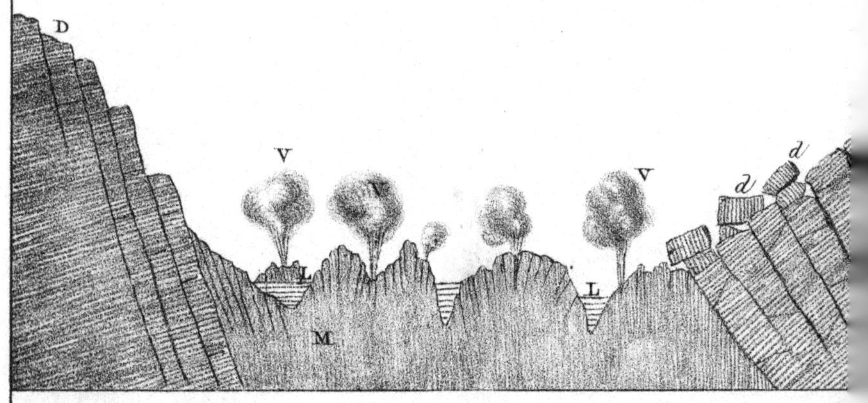

Fig: 1, 2. Disposition of the Serpentine & Compact Limestone, in the Environs of Mont

Fig: 3 _ Indication of the disposition of the Lagonis of Monte Cerboli

Fig. 4 _ Magnesite of Castellamonte near Turin.

G Scharf Lithog:

London. Publis

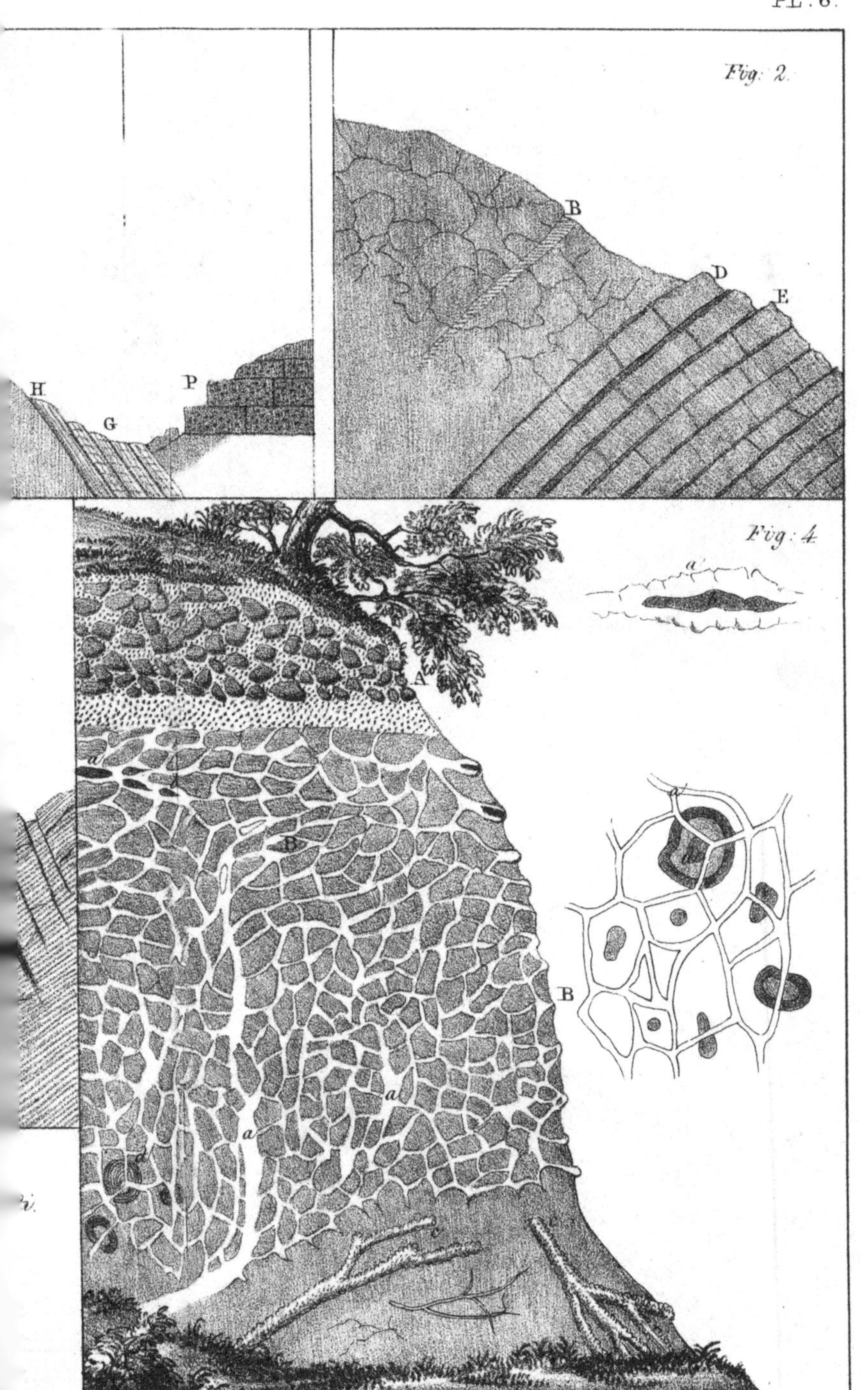

Printed by C. Hullmandel.

Phillips. 1823.

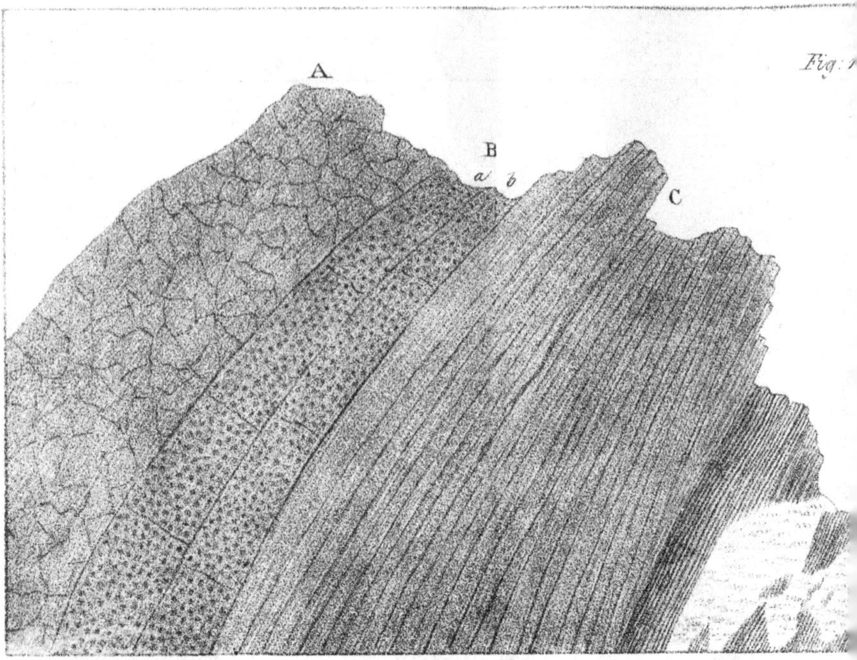

Fig: 1

A

B

a b

C

Relative position of the Serpentine A, *of the Diallage Rock* B *of the Jasper* C, *the Con*

Fig. 3

A

B

C

D

G. Scharf Lithog:

MONTE FERRATO ON THE N.W OF FLORENCE. T

Published b

Fig: 2.

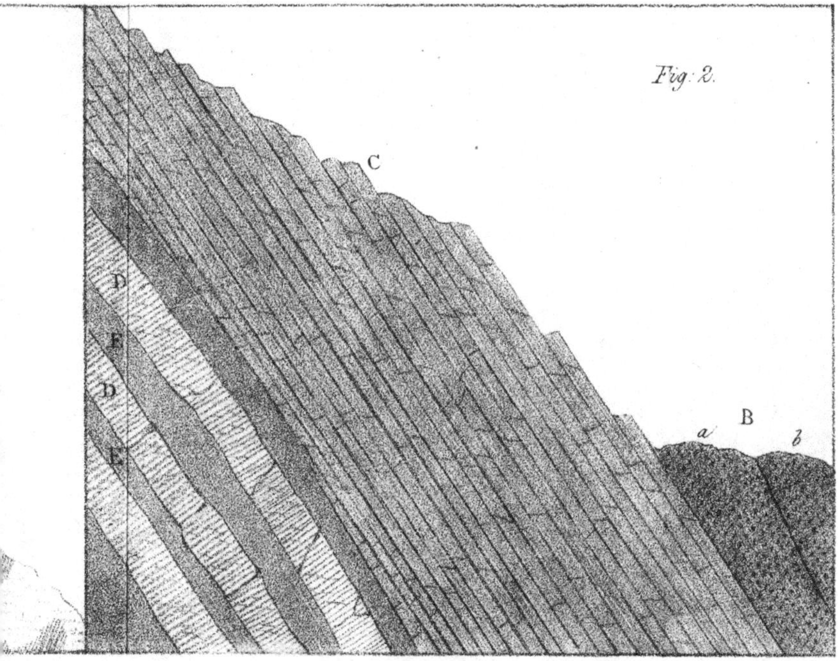

imestone D & the Marly Schist E, in the Valley of Cravignola, on the N.N.W of Spezia.

Fig: 4

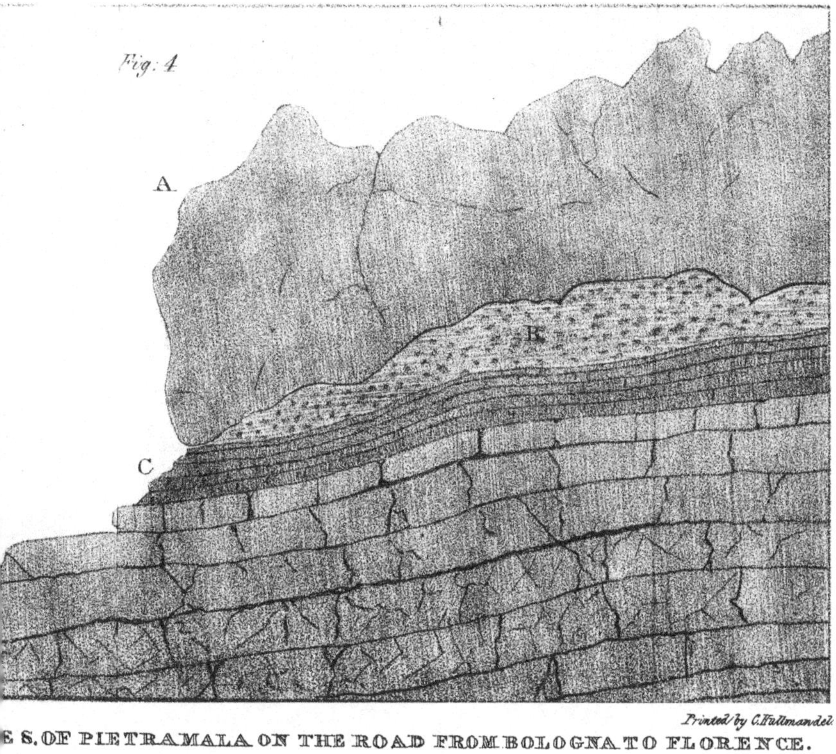

Printed by C. Hullmandel.

E S. OF PIETRAMALA ON THE ROAD FROM BOLOGNA TO FLORENCE.

hillips 1824

On the relative position of the Serpentines (Ophiolites), Diallage Rocks (Euphotides), Jasper, &c. in some parts of the Apennines; by Alexander Brongniart, *Member of the Royal Academy of Sciences, &c. &c.*

Read at the Royal Academy of Sciences, December 6, 1820.

(Annales des Mines for 1821).

Introduction.

IT is now very generally agreed that the end of positive Geology is to be able to understand, as exactly and completely as possible, the nature and structure of the crust of the globe, and to discover if general and constant laws have governed this structure.

The precise determination of the rocks, minerals, and fossil organic bodies constituting the different portions of this crust, and lastly that of their relative position, are the principal and perhaps only means that can enable us to acquire this knowledge, and lead us to the general, interesting and useful results that it promises. All geologists are agreed on these three propositions, and all endeavour to discover by different means the order of superposition of formations and the rocks composing them; but all are not yet equally agreed on the utility on determining previously, and independently of every other consideration, the rocks entering into the structure of formations. Some, without altogether rejecting this mode of considering them, attach very little importance to it; others, confounding it with the state of relative positions (positions respectives), desire that mineralogical descriptions should constantly accompany those of the formations. We have for a long time exerted our efforts to prove that these two subjects ought to be separately treated, that their confusion is necessarily carried into the science and retards the progress of it; we have with pleasure seen these principles admitted by many foreign and French geologists, and it is to be remarked that those even who do not desire explicitly to recognise them are compelled to submit to them, as it were in spite of themselves, in their geological observations. Those that I am about to bring forward will again prove, at least I venture the hope, the necessity of this distinction. It will be seen that it is very difficult, often even impossible, to be able to determine the relative position of formations composed of particular rocks and fossil organic bodies, if the differences of these rocks and fossils have not previously been made known with precision.

Among the rocks whose relative position in the crust of the globe are either obscure or little known, are reckoned certain ophiolites* (ophiolites), or rocks with a serpentine base, diallage rocks, and even jaspers; notwithstanding the labours and numerous travels for some time undertaken by celebrated geologists, the knowledge of the position of these rocks has remained either incomplete or uncertain; and I even venture to say, that in many cases, a very false idea has been formed of it; this state of uncertainty was principally owing to three causes:

1st. To the rarity of one of these rocks, the Jasper.

* I have given the characters of these composed rocks and those of their varieties in my essay on the mineralogical classification of mixed rocks, inserted in the Journal des Mines, vol. xxxiv. No. 199, July 1813. It is found translated into German, in the work entitled, "Taschenbuch für die gesammte Mineralogie;" by H. C. Leonhard, 9th year, page 378; and into Italian, in M. Moretti's work: "Classificazione delle Rocce. & Milano, 1814." M. de Bonnard has inserted the whole with some modifications and additions, which I am disposed to admit, in the 2d edition of the "Nouveau Dictionnaire d'histoire naturelle," at the word—roche.

2dly. To the confusion that has reigned, from the want of a good mineralogical determination, between the rocks with a serpentine base of different formations.

3dly. To the structure of the serpentines (ophiolites), in mass, which often occur as isolated mountains without distinct stratification, without foreign characteristic bodies, &c.

It certainly required this union of unfavorable circumstances to conceal the true position of these rocks from the observation of geologists distinguished for their science, their activity, and their works, who have before me travelled over that part of the Apennines, where I have with certainty observed the position of these rocks. I have at the same time acquired new proofs of the presence of rocks as perfectly crystallized as granite, placed on aggregate rocks as coarse as sandstone, on rocks perfectly resembling those which, in the same canton, contain organic remains. The facts I am about to expose will prove these two results, and the quotations I shall bring forward will shew that they were not only neither well known nor generally admitted, but that persons have often been deceived with respect to them.

Article I.

Enumeration and designation of the principal rocks composing that part of the Apennines which forms the subject of this memoir.

Those parts of the Apennines which are situated between Genoa and the north of Florence, and the environs of Sienna, not comprising the hills that border the Mediterranean, present three kinds of principal formations. which we shall designate by the following names, without considering, in this enumeration, the order in which they occur.

1st. The sandy, marly and shelly formation (terrain sablo-marneux-coquillier), or tertiary formation of Brocchi and almost all geologists.

2dly. The calcareous sandstone formation (terrain calcareo-psammitique; the arenaria, &c. of the Italian geologists.

3dly. The serpentine formation (terrain ophiolitique).

I mention the tertiary formation in this communication solely to shew that it is not forgotten; but as it bears no direct relation to the serpentine formation, the principal object of this paper, I shall not here speak of it.

There are some other rocks or formations, either subordinate to these, or in an independent position, such as the gypsum and saline formations. Those of marble or crystalline limestone, which I equally pass over in silence, either because they are foreign to my subject, or because I am not exactly acquainted with their relations, or lastly, because I have nothing more to say concerning them than has been published by Italian naturalists.

I confine myself therefore to the examination of the relative position of the serpentine with the calcareo-sandstone formations, and I ought, consistently with the principles I have above laid down, previously to determine, with the greatest possible exactitude, the rocks composing these formations.

1. The calcareo-sandstone forms, in this part of the Apennines, the principal mass of the mountains, and may be considered as the base or fundamental formation. It is composed of the following rocks.

a. A calcareous micaceous sandstone; that is to say, a rock apparently arenaceous, but solid and even compact, and hard enough to strike fire with a stroke of the hammer; more or less mixed with mica, and often traversed by veins of calcareous spar; its predominant colour is blueish grey with a crust of reddish brown (pietra serena of the Florentines). (Mountain of Fiesole near Florence; Oneille; Barigazzo.)

b. A micaceous sandstone (macigno and bardellone, Brocchi), passing into schistose sandstone and even a spangled (pailleté) argillaceous slate alternating with them (Doccia; Arezzo; Fiesole).

These two rocks differ but little from each other.

c. A very fine grained compact limestone, easy to break, with a conchoidal fracture, sometimes a little scaly; of an ash or bluish grey colour, with veins of calcareous spar.

(At Rochetta de la Spezia; at Doccia near Florence; at Pietramala, &c.)

d. A marly schist, sometimes rather solid, but extremely fissile and dividing in the manner of slates (between Barigazzo and the Col of Bosco-Lungo, road from Modena to Pistoia), sometimes of a yellowish brown colour, of a dull and even earthy aspect, and resembling marl; often even so fissile and disunited, that it is impossible to obtain a specimen fit for shewing its characters. (At Rochetta de la Spezia, on the borders of Cravignola). This schist passes into a dull argillaceous slate containing a little mica, into spangled (pailleté) argillaceous slate containing more mica, without from this circumstance ceasing to be marly, that is, ceasing to effervesce with nitric acid; in this respect very different from ancient argillaceous schist, which is dull, and does not contain lime.

Such are the rocks of a calcareous, arenaceous, and schistose nature that occur most abundantly in those parts of the Apeninnes I have mentioned. They alternate with each other without any order, often many times in a short space; they pass into each other by insensible gradations; they form entire and very elevated mountains, chains of hills and mountains of great extent, and offer some peculiarities in their structure, which I shall make known when I describe the places where I have observed the positions that constitute the principal object of this memoir.

2. The ophiolitic or serpentine formation is composed, in these same districts, of the following rocks, forming its principal and essential parts.

a. Asbestiferous greenish serpentine, brown serpentine with diallage, and common serpentine.

Throughout the whole valley of the Magra, and the Vara; in the environs of Prato, on the north of Florence; on the south, at Imprunetta; near Pietra-mala, on the road from Florence to Bologna; to the north of Genoa, at Mont

Ramazzio, and probably on the coast from Genoa to Inurea, &c.

b. Diallage rock (Euphotide of Haüy, Granitone of the Italians, Gabbro of M. von Buch). A rock composed, in the cantons to which I confine my observations, of compact or sublamellar felspar, of greenish serpentine, of schiller spar (diallage metalloïde), and sometimes perhaps of quartz.

On the western shore from Genoa to Savona; the felspar is granular, with quartz and more talc than diallage.

At Voraggio; the felspar is granular, with diallage or talc of a dull green colour.

At Figline, on the north of Florence; compact waxy felspar accompanied by lamellar blueish grey felspar, with diallage and a little green serpentine.

In the valley of Suvero and Cravignola, to the north of Brugnato, &c.

3. The Jasper formation (terrain jaspique) composed of extensive beds or mountains of jasper, subdivided into strata or even extremely multiplied and parallel laminæ, sometimes red with a dull fracture and coarse paste (at Prato; at Pietramala), sometimes red, of a finer paste, with either violet or dull green zones. It resembles the ribbon jasper of Siberia, though it does not possess either its hardness, its beautiful colours, or its fine grain, and it differs still more from it by perfect infusibility (at Montenero near Rochetta).*

There are two jaspers. The green and translucent part, which possesses a little of the exterior characters of petrosilex, is infusible with the blowpipe; the red part is equally so, but it becomes discoloured, and the surface of the scales becomes slightly shining.

This formation contains, as a subordinate rock, beds of altered red jasper, passing into red whetstone schist, and, as casual minerals, black oxide of manganese, compact and

* Dr. MacCulloch has given an account and synopsis of jasper rock, in his geological classification of rocks, division, 'Occasional rocks;' he states, "that it occurs indiscriminately both in the primary and secondary classes." (p. 539). (Trans.)

very hard, mixed with crystalline quartz and brown ochre or umber earth, which appears to be an alteration either of jasper or of manganesiferous red schist.

I shall not enter further into details respecting the rocks composing these formations; many of them have been well described by the Italian geologists, Messrs. Viviani, Bardi, Brocchi, &c. and by the French geologists, Messrs. Faujas, Cordier, &c.

It is moreover for naturalists who live on the spot deeply to study these details, that require time and frequent visits to the same canton. I do not therefore pretend to have completed this description, but to have sufficiently studied it for my object.

I ought now to shew the manner in which these rocks and formations occur together.

Article II.

Disposition of these rocks with regard to each other.

Sect. I. *Directly and closely observed.*

I have recognised these relations in three principal places, separated, in a right line, more than thirty leagues from each other; and this positive knowledge being acquired, I have been afforded the means of recognising the same mode of occurrence in places where it was not so evident, and to content myself with analogies, the value of which I now appreciate, in order to apply it to other foundations in which this disposition is shewn in an incomplete manner.

The first place where the relative position of the three formations is shewn in a perfectly clear manner, is the small valley of Cravignola, leading from Borghetto to the village of Rochetta; this village is situated at about 15 kilometres [$11\frac{1}{4}$ miles] to the north of Borghetto and Brugnato, and at 56 kilometres [about $42\frac{1}{2}$ miles] in a direct line to the N.N.E. of Spezia.

Following the bed of the torrent of Cravignola, a gorge is

entered that cuts the foot of two mountains, and which consequently exposes their structure.

That on the right bank which is the lowest, and which is represented almost entirely (fig. 1. pl. 5) shows the succession of the following rocks, inclined from N. to S., and covering each other. These rocks are, proceeding from the highest to the lowest,

1st. Green serpentine with diallage homogeneous, but much broken, and a petrosiliceous serpentine. The latter is compact, contains whitish petrosilex, disposed in irregular spots passing into little veins in the green paste. It contains cromate of iron, some pyritous points, and very little diallage.

2dly. Diallage rock (or granitone of the Italians) in thick beds, of tolerably regular size, and situated beneath the serpentine, (see fig. 1. at B.). This diallage rock is composed of white and greenish felspar or petrosilex, and shining diallage in very large plates. It offers some varieties, the principal of which are: *a.* serpentine diallage rock passing into the preceding rock, and only differing from it by more silex and the presence of diallage in large plates. *b.* calcareous reddish diallage rock of a greenish petro-silex paste with thin winding plates of steatite or greenish talc, and numerous red spots of sublamellar carbonate of lime: diallage is rare in it.

The upper beds of diallage rock, those that immediately follow the serpentine, are of a green paste, which colour is seen to be owing to serpentine. The inferior beds offer a mixture of a deep green, a pale green, a greyish, a white and red colour. The red part is not jasper, but as M. Viviani has observed, it is a red lamellar carbonate of lime. Diallage is there less abundant, and the rock is traversed by veins oi calcareous spar.

3dly. The rock that immediately follows it (at C) is a jasper, generally red but sometimes striped or zoned with a violet and greenish colour. This jasper is very splintery, at least on the superficies of the formation; it is cleanly stratified in beds more or less thick, rarely attaining two decimetres [about 8 in.], and often reduced into thin strata, and almost laminæ of from three to four centimetres [little more than 1 in.] thick; these beds, highly inclined from N.E. to S.W. or nearly so, are generally straight, parallel to each other, of greater thickness at the upper portions near the diallage rock; very thin and then a little sinuous below near the cape terminating the mountain. They are perfectly distinct, being uncovered throughout a great part of their extent, and concealed only in a few points by some patches of turf.

This mountain, which is rounded while mounting the torrent of Suvero, beyond the village of Rochetta, shews on that side absolutely the same rocks, and in the same position. It is seen, as well as in the portion I have just described, that the diallage rock of a red paste is always that which touches the jasper; which seems to indicate that it owes its colour to the same cause as that rock, and that it is very nearly of the same epoch of formation as this jasper.

The mountain (fig. 2), on the left bank of the torrent of Cravignola, bears the name of Montenero, and appears to be the foot or base of that which was shewn us by the name Mont Silva. It is not less steep than that on the right side, and being like it perpendicular at its base, it shews clearly its interior structure, and affords precisely the same series of rocks. The jasper, which is here a little more compact, and the same varieties of diallage rock, occur in the same positions.

But continuing to mount the torrent of Cravignola, and penetrating as it were into the interior of the mountain, we ought to be able to see, in consequence of the inclination of the beds, the rocks situated beneath the jasper, and consequently beneath the diallage rocks and serpentines; and, in fact the jasper suddenly ends opposite the confluence of the Suvero and Cravignola, and we see:

4thly. Alternating beds of a soft and very friable schistose rock (E), and a compact limestone (D), succeeding the jasper without interruption, and whose stratification is entirely parallel to it, as represented fig. 1. pl. 1. The friable schistose rock is not a primitive argillaceous schist, in the precise acceptation of that term, nor even an argillaceous schist, as it might at first sight be considered: for argillaceous schist, and above all that considered as primitive or even transition, is generally hard, of a fine and close grain, and often shining; if it contains mica, it is as it were dissolved and not in distinct plates. The schistose rock that occurs here beneath the jasper, possesses nearly the opposite characters, it is so earthy and easily broken, that a specimen of a certain size cannot be obtained. It is either of a dull yellowish, greyish, or even nearly blackish colour. The mica when met with, which is but very rarely, is seen disseminated in small spangles, difficult to be observed, and that which completes the series of its differences, is that it effervesces strongly with nitric acid.

This rock, when homogenous, is then a greyish marly schist, such as I have characterized it in my mineralogy, and when it contains mica, it is exactly referable to that which I have elsewhere * named dull micaceous argillaceous slate (phillade micacée terne).

This marly schist passes into compact limestone by insensible gradations, and alternates with beds of a fine compact limestone, of a smoke grey colour, and of a clean conchoidal fracture, traversed by numerous veins of calcareous spar.

The immediate and evident superposition cannot then here be doubted of the serpentine, of the diallage rock, rocks so clearly crystallized, and jasper on calcareous rocks which offer all the characters of a deposit (d'un sediment).

Before I quit this example and the place that furnished it, I ought to state some other facts that will complete its history.

While mounting up the Montenero or the mountain of Silva, and traversing the Col of Beverone, in order to pass by Garbuglaria into a valley which joins the valley of the Varra at Madrignano, the jasper formation is constantly on the right, that is towards the East, here rising towards the

* Mineralogical classification of mixed rocks, Journal des Mines, vol. xxxiv. p. 5. &c.

summit of the mountain, and in which the compact oxide of manganese is worked, that apparently occurs in disseminated nodules. This jasper is most generally red, and contains portions of agate; it is traversed by numerous veins of white quartz, which present cavities lined with crystallized quartz. When the Col de Beverone is passed, the jasper is seen on the left, towards the summits of the mountains, always resting upon the limestone and marly schist, shewing its existence by rolled fragments on parts of these mountains, and at a distance by its occurrence on thick beds, and by the reddish colour of these beds. The superposition of these two rocks could not be discovered from this obscure disposition; but being known, the analogous positions are easily traced here. Yet, as the serpentines do not rise to this height, so many rocks are not found, and it is only in descending from Beverone towards Madignano, that they are again discovered in the same relative position. While mounting towards the summit of Mont Silva, the jasper is on the right or East, and the limestone and marly schist on the left or West, rising from beneath the jasper; and as this calcareous formation occurs throughout a considerable extent, all its peculiarities may easily be studied.

The mixture of marly schist and compact limestone is seen to be still more frequent and complete here, than at the base of the mountain. This portion of the calcareous rock mixed with clay appearing to have collapsed more at the general desiccation of its beds, than the pure compact limestone, open clefts have resulted, which have been penetrated by marly schist. The latter being removed by some cause which it does not belong to my subject to search for, the lower portions of the limestone beds have remained divided into a multitude of prisms perpendicular to the plane of the beds, and separated from each other by open clefts. This disposition, remarked by Targioni and the Italian geologists, has caused the name of alberese costellino to be given to it, i. e. knifestone, because the ridges of the prisms are often very sharp. If the calcareous rocks affording these prisms were cut near their base, and perpendicular to their sides, the section resulting from it would offer that ruinous aspect that characterizes the prisms named the ruin-shaped marble of Florence.

This observation ought not to be considered as the description of an isolated fact, nor as the explanation of ruin-formed marbles. It is united to the geological history of these mountains, leading us to determine the geological position of the Florence marble, and reciprocally to establish in a more certain manner that of the calcareous formation which supports the serpentines, diallage rocks, and jaspers. We shall find this ruin-formed limestone in places far distant from this last, and the environs of Florence, in a geographical position, which bears the greatest analogy to that of the formation we are describing.

The second example, taken like the first from the northern side of the Apennines, is the least complete; we have seen but two different rocks in evident superposition; the position of the third must be presumed from analogy.

This second example occurs on the N.W. of Florence, near the small town of Prato, and to the W. of the village of Figline, and on the mountain named Monte-Ferrato. I had the great advantage of being conducted by M. Nesti, and Count Bardi, director of the Florence cabinet, who has published a very good memoir on the diallage rocks and jaspers on Monte Ferrato.

This mountain, examined on its southern side, presents a rounded summit, of little elevation, composed of serpentine and diallage, without any distinct stratification, but disposed on the contrary, in irregular masses as it were, separated from each other, yet without really being so, and traversed by veins of asbestus. It is on the summit of this mass of serpentine that M. Brocchi believes he has found and recognised prehnite.

Beneath this mass, at nearly a third of the mountain's height, and still on the southern side, the jasper appears in the section of a ravine (pl. 1. fig. 3), composed of coarse red jasper, mixed with a little greenish semi-transparent jasper. It is sometimes in thin laminæ, sometimes in thick strata, but always so splintery that a solid specimen of the size of an egg cannot be obtained. These very numerous strata, all parallel to each other, are highly inclined, and dip to the E.N.E. under the preceding serpentine. But it has been imagined that at the point of contact of the jasper and serpentine, the latter rock is altered; it becomes whitish, friable, granular, and has some appearance of an altered or imperfect diallage rock. Beneath the jasper, a portion of the mountain is passed over covered by vegetable soil, pasturage, and woods, which affords no opportunity of any where seeing the inferior rocks exposed, or in their primitive position. But large blocks of limestone (D), and even heads of beds are from time to time met with; this fine compact limestone, of a smoke grey colour, possesses all the characters of that of Rochetta, and although I have not seen it evidently beneath the jasper and serpentine, it may be presumed that it is, as at Rochetta, placed under these rocks; analogy indicates it, and no observation proves the contrary.

On this side the mountain the diallage rock is wanting; but on its northern side and towards its base, though still at an elevated situation, a thick mass of this beautiful rock is worked, which is here very hard. It contains, besides compact felspar that constitutes its base, a sublaminated felspar possessing a slight violet tint, and melting before the blowpipe into white enamel. Its slightly fat appearance causes it at first sight to be taken for quartz.

Large and good millstones are made of it.

Descending from these quarries into the plain, we find not the jasper bed, it is probably too thin, and appears to have been reduced in thickness as much as the bed of serpentine has augmented, but small fragments of jasper disseminated in the soil, that indicate its place, and lower down the fine compact limestone of a smoke grey colour.*

* Count Bardi had remarked the position of the serpentine on the jasper, and had stated it in a memoir published in the 2d volume of the Memoirs of the Florence Institute, entitled, Osserv. miner. sopra alcuni luoghi adjacenti alle pianura di Prato. He says, page 20 of this memoir, " The gabbro (serpentine) immediately rests on the jasper, and reaches

Let us now pass on to the third example of the superposition of the serpentine jasper formation on the calcareo-marly formation. This seems the most complete: all the rocks appear clearly above each other. It is no longer at the foot of the Apennines, but placed on the crest of that chain of mountains. It occurs in a place of easy access, I would say more, in a place frequented by all travellers, and consequently by all the geologists who pass from Florence to Bologna.

As I visited this place some time after those preceding, I then considered that the superposition that I had remarked at Rochetta, and regarded as a fact not much known, ought to be described in geological works; and it was with astonishment, and even yet mistrusting the exactitude of my researches, that I presume that it had escaped them, or rather that, seduced by false appearances, and led by a kind of prejudice on the antiquity of the diallage rocks and serpentines, they have not been willing to refer to this formation the rocks of a serpentine base that form the summit of the mountains.

It is to the south of Pietramala, on the side of the great road, close to a place celebrated from the hydrogen gas that is perpetually disengaged from the earth, and which is almost always lighted, that an example of this superposition occurs, as clear and more evident, if possible, than that of Rochetta: for here the rocks are nearly horizontal, as is seen pl. 5. fig. 4.

I should omit mentioning the rocks which are met with before the point is reached where the superposition of the serpentine formations begins to be clearly seen; not because

to the summit. Yet in other parts of the same mountains, the jasper rests on gabbro;" which would be equally possible, if these two rocks are, as it appears, of contemporaneous formation. But not being able to see this superposition in the visit made to this mountain by Count Bardi, M. Nesti, and myself, the former appeared to adopt my opinion on the superposition of the serpentine to the jasper.

He elsewhere remarks that pieces of compact limestone are found in these two rocks; but he regards them as produced by crystallization and of contemporaneous formation with the jasper and serpentine.

the history of these rocks would be without interest to science, but because I was unable to observe them with sufficient care, and as it is not necessary for my principal object.

I ought only to state that at Fontebuona, a col of the first line of Apennines on the side of Florence, on the road from that town to Bologna, the alternation occurs of the greyish fine compact limestone, already described at Rochetta, with the calcareous and micaceous sandstone, which is compact and schistose, an alternating rock not so clearly seen at Rochetta. This disposition occurs again at a place named the Maschere, and among the fragments of the sandstone that alternates with the limestone which is brought on the road; I found one, but only one, that contained some carbonaceous bodies.

Shortly afterwards, mounting to the place named lo Stale, Monte Carelli appears a reddish isolated knoll, composed of broken stones aggregated together, without apparent stratification, consequently of a true breccia, formed by the union of fragments of red jasper, serpentine, &c.*

It was, it may be said, announcing the near presence of

* This knoll, from its form, its isolation, its colours, and the mode of aggregation of the rock composing it, has been taken by Ferber for volcanic. He does not hesitate to say:

"The Monte-Traverso is formed of lavas and owes its origin to an ancient volcano." It certainly is not one in the general and received acceptation of that name; but its appearance is deceiving, and arriving on the spot without recollecting what Ferber had said, we were struck with the same idea, and we could not avoid saying that this hill bore a singular resemblance to the extinguished volcanoes we had seen.

Ferber is not the only person who has considered that a serpentine rock might be of volcanic origin.

Guettard had remarked blocks of serpentine between Loretto and Ancona in a formation he considered volcanic. The existence of the volcanic formation is far from being proved; but the idea of the association does not the less remain. Sir G. Mackenzie states that the volcanic amygdaloid beds of the mountain of Akkrefell, in Iceland, are traversed by veins of serpentine of more than a metre [about 3 ft. 3⅓ in.] in thickness. Lastly, M. Breislak does not find any reason for excluding magnesian rocks from volcanic products.

serpentine rock, diallage rock, and jasper, a notice to examine the mountains attentively. We mounted towards an elevated platform here forming the crest of the Apennines; the stratification was nearly horizontal, consequently we presumed that as we rose we quitted the inferior and arrived at the upper beds. In fact, after passing the col of lo Stale, on the side of Traversa, in a very hilly canton named, we were informed, Sasso di Castro, the succession of the following rocks begins to shew itself, continuing nearly from Maschere to beyond Covigliano, a short distance to the south of Pietramala. This succession of rocks becomes even more distinct the nearer Pietramala is approached, and appears well defined in the almost perpendicular mountains on the west of the road.

1st. The summit A, fig 4, of these mountains, which is precipitous, or with a very steep slope, cut by numerous furrows or deep ravines, and bristled by small peaks, or very pointed paps, is composed of hornblende serpentine (ophiolite amphiboleux), and especially of hornblende diallage rock (euphotide amphiboleuse), resembling greenstone, but which is too soft to belong to that class of rocks. These rocks pass, at the base of the mountain, into serpentine diallage rock (B), which here presents a very remarkable variety, which is that it possesses the amygdaloïd structure, and all the characters of rocks named amygdaloïds (variolite). The white spots appear to me to be petrosilex; they melt very easily before the blow-pipe, but with a very remarkable ebullition. I make a peculiar variety of it, by the name of amygdaloïd diallage rock (euphotide variolitique).

This rock varies in mineralogical structure, and I might long descant on its varieties, if the details had been of use to my subject. It is more or less thick, and enormous fallen masses occur at its foot.*

* Circumstances not allowing me to remain at Pietramala, I was unable to climb to the summit of this mountain in order to examine in place the varieties of the serpentine rock composing it: I studied them among the immense fallen masses which cover the sides, and which present

2dly. Beneath this diallage rock a red bed (C) is seen, of nearly equal thickness throughout its whole extent, almost horizontal, or very little inclined towards the North. It is jasper in thin and nearly parallel beds. It is principally red, yet is occasionally mixed with greenish zones.

enormous masses evidently detached from the unstratified but broken blackish summits of this small chain of mountains; they are so voluminous that, if sufficient attention is not paid to it, it might be thought that these masses of serpentine rocks were in place, and that they were beneath the limestone; it was a remark I made on the spot to my young companions, M. Bertrand-Geslin and my son. But M. Mesnard de la Groye has supplied what may be wanting, by communicating to me with a generous eagerness the specimens he collected on the summit of this mountain. They have shewn me that there occurred in these well characterised serpentines very remarkable veins of crystalline quartz, containing pyrites; as also a hornblende diallage rock, in which the petrosilex, the diallage, the hornblende, and even the serpentine are perfectly distinct; lastly a porphyritic rock, traversed by very singular calcareous veins, and which unites the serpentine formation to that of porphyries. Whether this serpentine rock is less abundant in serpentine than the others, as I suspect, or that it even passes into greenstone, it does not the less belong to what geologists call the second serpentine formation. M. von Buch, in the examples he gives of gabbro, i. e. of this formation, cites Covigliano; and in the MS section he has made and confided to me, he places a summit of serpentine in this spot.

The association of serpentine or steatite with hornblende is well recognised, and referred, like the serpentines, to the transition series. M. Stifft mentions, to the S.W. of Neubourg, a thick bed of steatite on an altered sediment basalt, accompanying greenstone, placed on transition limestone, near Herborn (Leonhard, tasch: 1808, p. 216.) M. Daubuisson also admits this association, and remarks that the passage from serpentine to hornblende is often insensible. M. de Bonnard equally brings forward, as an admitted fact near Hartzburg, in the Hartz, the passage of diallage rock into greenstone, by a diallagic greenstone, &c.

The rock on which M. Palassou has written so much, which he names Ophite, and which is actually a greenstone, passes into serpentine and hornblende diallage rock; it belongs, as he himself has remarked, and I have had occasion to observe near Pouzac, to the serpentine formation, and he insists that it rests on a secondary limestone.*

* Dr. MacCulloch remarks in his account of serpentine, (Classification of Rocks, p. 245), "That when the contact is that of hornblende rock with serpentine, a perfect gradation may sometimes be traced." He states two examples have been seen by him of serpentine in the secon-

It is so brittle and so broken by atmospheric phenomena, that its debris form long red slopes, which appear to cover the escarpments, and partly conceal the following rocks.

3dly. The formation that is immediately beneath it, occurs in beds having a stratification parallel to that of the jasper, and which apparently forms the very thick base of this elevated part of the Apennines; this formation I say, is principally composed of a fine and compact limestone (D), of a smoke grey colour, and conchoidal fracture, traversed by numerous veins of calcareous spar, and of a compact yellowish grey limestone without spathose veins. I here notice the principal and most abundant varieties; but others occur that I have not considered worth describing. Notwithstanding our research, we have not been able to discover any organic remains in these rocks.

Beneath this limestone, and alternating with it, occur the hard compact sandstone, and the schistose sandstone, which is here often very micaceous (F).

From the midst of this rock rises the hydrogen gas of Pietramala, and this position, it may, en passant, be remarked, is absolutely the same as that of the hydrogen gas of Barigazzo, on the road from Modena to Pistoia.

The col that is passed on the north of Pietramala is entirely composed of this same fine and compact limestone, and on the descent the same micaceous sandstone is met with in thick beds inclined towards the north.

Here then occur nearly the same rocks as at Rochetta, in the same order of superposition. The varieties of very little importance presented by these rocks, are those that ought to be expected throughout the earth between bodies

dary class. "In both these cases, veins of trap pass through strata of secondary limestone, and where the vein is in contact with the limestone, it changes its character and becomes a serpentine, while it contains in those parts the minerals usually found in that rock, namely asbestos and steatite. The limestone, at the planes of contact, also contains steatite; and thus a species of regular gradation becomes established between the trap and limestone. The gradation from the serpentine to the trap, within the vein, is perfect and insensible." (Trans.)

of the same nature observed at some distance from each other. But here we have more than at Rochetta, the alternation of the limestone and the micaceous sandstone, of a coarsely aggregated rock with one of a fine deposit (sédiment fin), and the whole beneath rocks, the complete though confused crystallization of which indicates an entire previous solution. Such is the third example I have to adduce as to the evident superposition of the serpentines and diallage rocks, crystalline rocks, on compact limestone and micaceous sandstone, rocks of sediment * and aggregation. If any doubts remain on the alternation of these two last rocks, and of the superposition of the limestone on the sandstone, they will be removed by the facts that I shall hereafter expose, when I endeavour to determine the epoch of formation to which these rocks may be referred, and by the exact coincidence of my observations on this last alternation with that of M. von Buch.

§ II. Position of the serpentines in other parts of the Apennines, determined by analogy.

I shall not pass in review, with reference to the observations I have just made, all serpentines and diallage rocks, in order to compare that which is known respecting their position with what we have learned respecting that of the Apennines: this enumeration would carry me too far away from the principal object of my work; but I ought nevertheless to examine if the position of any well known serpentines, and especially those in other parts of the Apennines, presents a disposition contrary to that which I have above made known, or if they are not rather the same.

I have seen the same serpentine formation, that is to say, the association of serpentine, containing diallage with diallage rock, near Monte Cerboli in the Volterranais; at Mont Ramazzo near Genoa; at the Bocchetta, on the north

* I have considered it proper to use the author's own word of "sediment," here and elsewhere in this memoir, as I could not otherwise so well convey his meaning. (Trans.)

of Genoa; at Castellamonte and Baldissero, near Turin; and notwithstanding the difficulty of recognising, or even the impossibility of seeing the rock beneath these serpentines; notwithstanding the differences they present, I consider that their position may be presumed the same as that of the serpentines of Rochetta, Pietramala, and Monteferrato.

To the south of Volterra, and beyond Pomerance in Tuscany, on the way to the lagonis of Monte Cerboli, a high hill is traversed, of some extent, bearing the name of Poggio del Gabbro,* (fig. 1.), and which is entirely composed of diallage serpentine. On the ascent we find a blackish compact limestone at the foot of the hill, afterwards rolled pebbles of serpentine and jasper, then serpentine in a thick mass. Descending from the col of Monte-Cerboli, towards the S.S.E. a bed of gypsum is first met with, which appears to rest against the serpentine, and covered in one spot by a conglomerate in thick and nearly horizontal beds, composed of all kinds of rocks, and especially fragments of serpentine; continuing to descend in order to reach the valley of the lagonis we quit, after passing Monte Cerboli (fig. 2), the serpentine, and meet with diallage rock, and afterwards limestone, the numerous and regular beds of which, separated by beds of marly limestone, dip under the mountain, and consequently under the diallage rock.

In this limestone are situated what are called the Lagonis of Tuscany (fig. 3). Their description and the account of their position, are foreign to my subject; I nevertheless cannot avoid remarking, that the vapours of boiling water, which rise with great violence from the clefts of this limestone, and which contain boracic acid among the substances that they carry up with them, take their rise beneath this rock, or at least in it. As not any mineral is known, in the mass of this limestone, either here nor elsewhere, which could give rise to phenomena that are so powerful, so extensive, and so general throughout all this country, nor to

* A new proof that Gabbro is the Italian name for serpentine, and not diallage rock, which is generally called granitone.

the various matter carried up by these vapours, I presume that they originate beneath this limestone, and that consequently the newest rock in which the focus of these phenomena can be placed, would belong so the transition formation.

Monte Ramazzo, a mountain on the N.W. of Genoa, which is a continuation of that of Guardia, is composed of diallage serpentine, containing copper pyrites, and gives rise to works of sulphate of magnesia, described by Messrs. Faujas, Moyon, Viviani, Cordier, &c. of amygdaloïd serpentine, rare, certainly, but perfectly resembling that of Pietramala, and of calcareous schist (calschiste) passing into steaschist. The formation on which the serpentine is placed, is here different from the calcareo-sandstone formation that I have observed in the places cited above. It appears to be wanting, and the serpentine is placed immediately upon a transition, and perhaps even primordial calcareo-talcose formation, very different from the preceding. It possesses all the characters of ancient rocks; the limestone is almost lamellar; it is mixed in thin, tortuous beds; and is as it were dissolved with the shining slate and steaschist; but the serpentine is not covered by any other rock, consequently nothing shews it to be inferior to the rocks that I have referred to the Alpine formation.*

Diallage serpentine also occurs at the pass of the Bochetta, and which, situated on the north of Genoa, forms part of the crest of the Apennines, in this portion of Liguria. It is very difficult to observe its position: it appears to be the

* M. Faujas (Annales du Museum, t. viii. p. 313), states, that in the torrent of Charavagne, serpentine is seen united to limestone by veins of spathose carbonate of lime, and he has remarked the amygdaloïdal diallage rock (euphotide variolitique) which he describes by the name of Variolite à base de serpentine.

M. Holland (Annales de Chimie et de Physique, t. iv. p. 427,) has also given a description of this mountain, and of the manufacture of sulphate of magnesia there established. He states that primitive schist is undoubtedly the base of the serpentine formation, which rests on it in considerable mass, and in an unconformable position.

same as that of Monte Ramazzo, i. e. that this serpentine occurs with, or even in the steaschist and calcareous slate, composing these mountains: for, nothing that I was able to see—nothing brought forward by de Saussure, M. de Humboldt, in the notes he communicated to me, and M. Cordier, prove a contrary disposition. But I abandon this point as yet obscure, in order to throw new light on a position little or imperfectly understood, by determining that of the rock or marble celebrated in the arts by the name of Vert de mer, and which I have named Ophicalce veinee, in my mineralogical classification of rocks.

This rock, which is well exposed to the E. of the village of Lavezara, appears to form part of a mountain composed of serpentine and steaschist, i. e. of the serpentine formation.

If we recollect that at Rochetta, beneath the serpentine, and immediately above the jasper, we recognised a calcareous diallage rock composed of green talc, whitish petrosilex, a little diallage, and red spots that were in a great measure calcareous; that this rock is moreover traversed by numerous veins of calcareous spar, and if we compare this diallage rock with the Lavezara marble, we shall find the same characters of structure and nearly the same composition, and even the same colours, i. e. white spathose limestone, red steatite limestone, and green talc. Only here, the limestone is most abundant and the felspar appears to be wanting, for I cannot affirm that it actually is so. Perhaps from researches that a traveller cannot undertake, it may be discovered in some parts of this rock. Led by analogy, which may, in geology, be regarded as a sure guide, at least in the same canton or system of mountains, I do not doubt but that the marble or ophicalce veinée of Lavezara, is a mineralogical modification of the calcareous diallage rock of Rochetta, and that it has exactly the same position; consequently that this rock is, like the diallage rock beneath the serpentine, and above the jasper and calcareo-sandstone formation, and that, far from being a primitive rock, as has been said, it belongs to a later formation, that of the

Alpine limestone, and probably more recent than the transition rocks mentioned at the commencement of this memoir.

The hills of Castellamonte and Baldissero, at the foot of the eastern side of the Alps, eight leagues to the N.W. of Turin, present the serpentine formation in a state of alteration that seems to remove it entirely from the rocks I have just described or mentioned.

But when we do not attach ourselves too much to mineralogical details; when we place ourselves as it were at a distance, so as to cause these details to disappear, and to see the whole together, the serpentine formation is recognised in these hills with all its essential circumstances. Diallage serpentine occurs as the predominant rock, especially at Baldissero, but is rarely solid; it is even extremely altered, traversed by a multitude of veins of magnesite (B). The diallage rock, instead of being a bed in the midst of, or beneath the rock, appears to be disseminated in nodules, (*d*) pure at their centres, but more and more altered from the centre to the surface.

The jasper is still found, but it passes into hornstone; it is yellowish or greenish, and instead of occurring in beds extended beneath the diallage rock, it forms nodular veins (*c*), that are irregularly ramified in the midst of the magnesian serpentine. It is not abundant, at least in the places I have visited; but it appears to have been replaced by every variety of opal, which is disposed in small irregular plates (*a*), mammilated on the surface, in the midst of magnesite veins. Such are the characteristic features of these mountains which I have no intention of describing; for, to make them known, more time must be devoted to it than I have given, and details must be entered into that would too much augment the extent of this memoir.

This formation is only covered by one of rolled rocks (A), to which I shall elsewhere return: this is an interesting point to remark, because it presents the commencement of the tertiary rocks (sediment superieur), that cover so large a portion of Italy from Turin to the extremity of Calabria.

Thus the serpentine formation is no more covered here than elsewhere. The rock on which it rests is certainly not seen; but the analogy of its structure and composition with those I have described, lead me to presume that it belongs to the same epoch of formation, and that it is, like them, posterior to the alpine limestone, such as I have described it.

To these rocks of serpentine that I have visited, I consider myself able to add, as referable to the same epoch of formation :—

The serpentine of the mountain of Dragnon on the side of Sasseto, in eastern Liguria, described by M. Viviani; it is so near the position of Rochetta, and the characters given by this naturalist are so like those of the Rochetta serpentine, that I have no doubt but that they belong to the same formation:

The serpentine of the mountain of La Guardia, on the N. of Genoa, described by Saussure. I regret not being able to visit this position; for, from the description of Saussure, it would appear, that the jasper is here replaced by red slate, and that the grey alpine limestone, alternating with marly limestone, found immediately under the serpentine, is stratified unconformably to the calcareous slate, and the blackish limestone traversed by spathose veins, which appears near Genoa, and near the borders of the sea, and which affords, better than all the limestones of this canton, the characters of a transition rock. Now this discordance of stratification, is, if not a certain, at least a very probable indication of different epochs of formation.

Every thing then coincides in shewing that the serpentine formation of the Apennines, far from being beneath the transition schists, as some celebrated geologists have imagined; far from belonging to the primordial formation; far even from closely following and being either a last member of that formation, or one of the most ancient rocks of the transition series, as has been said by Messrs Von Buch, Faujas, Viviani, Cordier, Cortesi, Brocchi, and perhaps all geologists, is, on the contrary, one of the last rocks of that formation, if even it belongs to it, i. e. if it were wished absolutely to comprehend in the transition series the calcareous sandstone of the Apennines and the Alpine limestone, which is, as I shall endeavour to prove, of the same epoch as that of the Apennines, and perhaps even more ancient.

Article III.

Determination of the epoch to which the Sandy Limestone immediately beneath the Serpentines belongs.

It is not sufficient to have recognised, as clearly as may be desired, the order of the superposition of the rocks I have above described; it must now be determined, if possible, to which epoch of formation, i. e. to which of the great divisions of rocks comprising the crust of the globe, these should be referred, with which we are at present engaged, or at least the lowest of them, for they nearly determine the age of the others.

Geologists have generally admitted the three great divisions which are not volcanic, established by Werner and his numerous disciples: the primitive rocks; the transition rocks; and the secondary or rocks of sediment; but the latter, lately studied with more care, have shewn the necessity of a subdivision, and I consider myself to have been one of the first to establish it under the names of inferior, middle, and superior sediment rocks, giving each of these groups as precise limits as the state of science would permit.

I shall neither recall here the motives of these divisions, nor the characters of the formations they contain: these are known circumstances, developed in geological works, the details of which would too much lengthen this memoir. I shall content myself by calling to recollection, as a circumstance less known, and what is not perhaps admitted by all geologists, the limits that I have assigned these different sediment formations.

The inferior sediment formation extends from the last transition rocks to the graphite limestone inclusive. It contains the coal measures especially, the alpine limestone of the German geologists, and * the lias and mountain limestone of the English geologists.†

The middle sediment formation extends from the preceding limestone to above the chalk, and principally contains the compact, whitish, and oolitic limestone of the Jura, and the chalk. It is a very extensive formation, to which I shall hereafter have occasion to return in another memoir.‡

The superior sediment formation, also named tertiary formation, extends from the chalk exclusively, or from the plastic clay and lignites exclusively, to the surface of the earth, or rather to the last marine deposits of the ancient sea.

The formation beneath the serpentines may either be referred to a transition formation, or the inferior sediment formation. Whatever may be the opinion adopted, a new fact in geology must always be admitted, i. e. a transition formation very different by its characters from those generally referred to that series, or serpentines and diallage rocks of a new formation, since, in the second hypothesis, they would be at least posterior to the inferior sediment or Alpine formation.

The first idea that presents itself, that which a justly celebrated Italian geologist, M. Brocchi, has entertained

* I have inserted the word "and," conceiving that M. Brongniart does not confound the lias and mountain limestone. (Trans.)

† The inferior sediment formation of the author therefore consists of—old red sandstone? carboniferous, or mountain limestone, millstone grit, coal measures, magnesian limestone (Alpine limestone of the Germans), saliferous, or new red sandstone, and lias.

‡ The middle sediment formation therefore consists of the oolite formation, from the inferior oolite upwards, the beds between the oolites and chalk, such as the iron sand, weald clay, and green sand, and the chalk formation.

and published, is, that the rocks beneath the serpentines belong to the transition series. Geologists who have had numerous opportunities of seeing and studying these rocks, Messrs Von Buch, Buckland, &c. do not admit this determination, and regard them as much more modern.

I am disposed, notwithstanding the very respectable authority I have above cited, to place myself on the side of the latter geologists, and especially M. von Buch, and to refer these rocks to the inferior sediment formation, i. e. to a formation or collection of rocks which they much more resemble in all the characters they present, than those of the true transition class.

It is by comparing the rocks in question with those which are admitted by almost all geologists, some as transition rocks, others as inferior sediment or alpine rocks, that the solution of this question may be more surely determined.

But in order to render the comparison more perfect, the characters of the formation beneath the Apennine serpentine must be completed, by examining the rocks in other places than those I have mentioned, which, appearing to me of exactly the same formation, and being admitted as such by Italian geologists, present in their structure, in the bodies they contain, and in their position, characteristic peculiarities, that the positions of Rochetta, Prato, and even Pietramala, do not offer in so clear and complete a manner.

§ 1. *Identity of the rocks beneath the serpentines above described, with rocks in other parts of the Apennines.*

The first place I shall notice, because it is not far distant from those which have formed the principal subject of this memoir, because I could study it with care, and because it presents in a complete manner the rocks composing the calcareo-sandstone formation, in the park (parc) of Doccia di Sesto, to the N. of Florence, a village in which is situated the porcelain manufactory of the Marquis Ginori. The part of the mountain exposed to the S.W. shews the naked rock in many places. Oblique beds are there observed,

1st. Of a fine compact limestone, of a pale ash grey colour, of a conchoidal fracture, traversed by numerous calcareous spathose veins, and completely resembling that of Rochetta and Pietramala.

2ndly. Of a hard and micaceous calcareous sandstone, traversed by spathose veins, and entirely resembling that of Pietramala, Barigazzo, &c.

3dly. Of a dull marly argillaceous slate.

These three rocks alternate together, I do not say without real order, but without any order as yet understood; there is not as far as this any difference between this formation and that which is beneath the serpentines at Rochetta and Pietramala. The limestone is a rock common to the three points; the calcareous sandstone, common to Pietramala, and Doccia, establishes the resemblance of this latter place to Rochetta, where I have not seen the sandstone, and to Barigazzo, where I have not seen the limestone.

But there is here a peculiarity in this limestone that I have not observed in the other places, it is the presence of hornstone in numerous nodules, placed in the same line. This peculiarity seems greatly to remove this limestone from that which is commonly considered as belonging to the transition rocks.

The high hill of Fiesole, on the N.E. of Florence, forming, like that of Doccia, part of the first line of the Apennines on this side, is celebrated for the numerous quarries there worked, and which furnish the stones employed in all the works at Florence. It shews, from about a third of its height to the summit, a very solid micaceous and calcareous sandstone, of a greyish, blueish, and yellowish colour, in beds sometimes horizontal, sometimes highly inclined in different directions, but more particularly towards the north. This sandstone, completely resembling that of Doccia, Pietramala, Barigazzo, &c. alternates with beds, more or less thick, of yellowish micaceous argillaceous slate, and exposes fragments of brownish schistose sandstone, which have been sometimes taken for portions of vegetables. Now this rock being considered as greywacké, by Messrs. Von Buch, Brocchi, &c. consequently as a transition rock; being the same as that found in the park of Doccia, two leagues from Fiesole, in conformable stratification with the limestone with silex, establishes, upon great presumable evidence, notwithstanding the recent appearance given it by the presence of silex, that it belongs to the same formation as the Fiesole sandstone.

If in the mountain of Fiesole the limestone is not found in place alternating with the sandstone, numerous fragments of that rock are found at the foot of the mountain, indicating that it is not distant.

If we afterwards proceed to the other side of the valley of Ombrone, on the hill of Seravalle, a short distance west from Pistoia, a smoke-grey compact limestone is found, traversed by spathose veins resembling that of Doccia, Rochetta, and Pietramala; this limestone moreover contains small veins of spathose iron; it alternates with a brown marly limestone, schistose, but solid, just shewing some spangles of mica, in this resembling that of Rochetta, and with hard micaceous calcareous sandstones, and yellowish spangled clay slates, resembling those of Doccia, and only differing from those of Fiesole by the small thickness of the bed.

Lower down, i. e. still more west, and towards the sea, between Lucca and Massa-Rosa, above a very different limestone from the preceding, and of which, for that reason, I ought not to speak, beds of compact limestone are found, whitish or slightly yellowish grey, but fine grained, with a scaly fracture, traversed by spathose calcareous veins, and resembling, by these characters and nearly by the shade of its colour, those of Rochetta, Pietramala, Seravalle, and Doccia, and containing, like the last, hornstone in thin seams or nodules disposed in the same line. These circumstances already make it presumable, if even they do not entirely prove, that the limestone with silex is of the same formation as the calcarcous sandstones, the dull argillaceous slates, the smoke-grey compact limestones, &c. and that it is consequently inferior, like all these rocks, to the serpentine formation.

I might multiply citations and consequently resemblances; but those that I should add not affording any thing more striking than the preceding, I conceive that I have sufficiently made these rocks known, to be able now to compare them with those generally noticed; some under the name of transition rocks, others by those of Alpine limestone or inferior sediment rocks.

§ II. *Comparison of these rocks with the transition rocks most generally admitted as such.*

If in the first place we compare the rocks we have described with those that almost all geologists refer to the transition epoch, we shall find but very little resemblance between them.

We in fact see in the Apennine formations greyish rocks of pure compact limestone, but without any crystalline appearance in its paste, passing on the contrary to the schistose texture and marly state; sometimes arenaceous and micaceous rocks, always calcariferous and nearly marly, not containing any organic remains resembling those that are admitted to be found in the transition rocks, being very regularly stratified and often nearly horizontal, containing hornstone either in their mass, or in the rocks which occur with them in parallel and continuous stratification; not containing, in the numerous places where I have been able to observe them beneath the serpentines, any of the metals so common in the transition rocks, &c.

What relation, I say, can be found between these rocks and the transition formations of argillaceous schists, pure spangled clay slates, i. e. not calcareous, containing schistose jaspers and aluminous slate, alternating with the black sublamellar limestones almost always fetid, containing lead and zinc ores, anthracite, &c. and which is seen in England in the environs of Bristol,* in Wales, at Altenlead [Allenheads] in Northumberland, &c. with those of Norway described by M. von Buch, which are so well crystallised that one would be inclined to refer them to the primitive class, if the black limestones and aluminous slate, containing organic remains, did not form part of them.

If the granular sandstones of Clausthal bear, at first sight, some resemblance to certain sandstones of the Apennines, they differ from them much more by the presence of felspathic grains, to which they in part owe their granular structure, by numerous metallic and calcareous spathose veins which traverse them; and yet none of the sandstones or slates of the Hartz are calcareous, nor the argillaceous schist of Nägenthal near Altenau, nor the spangled yellowish argillaceous slate of Schalk near Schulenberg, which contains so many remains of entrochi, nor that of Rammelsberg, finally none of those in the Hartz which I have tried, effervesce; all the sandstone rocks of the Apennines are, on the contrary, very effervescent.

What I have said of the Hartz applies to the transition formations of Saxony, which, from their aspect, perhaps differ still more than these from the calcareo-sandstones of the Apennines.

The sublamellar black limestones of the environs of Namur, Mons,† &c. which all geologists refer to the ancient transition formation, have not any resemblance to the grey compact limestones of the Apennines.

The transition rocks which I have seen in France at Montchatou near Coutances, and which sufficiently resemble

* The author is in error with regard to the Bristol limestone, which is the medial or carboniferous limestone, and not the submedial or transition. The Northumberland is most probably the same; as may also the limestone the author mentions in Wales, at least if he means that most abundant in South Wales: foreign geologists generally refer this rock to the transition series, the limestone of which it certainly more resembles than that of the secondary. [Translator.]

† This limestone is also analogous to our carboniferous or mountain limestone. [Translator.]

those of Bristol;* those of the environs of Cherbourg, which I have described, and which are composed of yellowish and not effervescent argillaceous slate, clay slate, aluminous slate, felspathic and granitic rocks, &c. those of the environs of Angers, which principally consist of a spangled clay slate, containing organic remains, but not effervescent; all these rocks, I say, compared with the calcareo-sandstones of the Apennines, present numerous and striking differences.

Even in the Pyrenees, the black spangled slates with vegetable impressions, and the brownish and micaceous schistose sandstones of the port of Gavernec, which some geologists refer to the transition series, possess an exterior aspect, a colour, a general disposition which distinguishes the arenaceous and slaty rocks of the Apennines; and what is still more remarkable is, that in the Pyrenees, notwithstanding the vicinity of these rocks and the calcareous formation, none of them effervesce, whilst all those of the Apennines that may be compared to them are mixed with carbonate of lime.

If we draw nearer the countries which contain the subject of our observations, taking as an object of comparison the Tarentaise in the chain of the Alps, a transition country rendered classic by the learned description M. Brochant has given of it, we find, notwithstanding the proximity of the places, but very few points of resemblance. The general mass of rocks in the Tarentaise is crystallised or granular; the striped schist of la Magdeleine, the base of which is compact, is traversed by crystalline limestone in every direction; the only limestone of a compact appearance there noticed, is that of the Bonhomme, which, by its yellowish white colour, by the fineness of its grain, which renders it almost translucent, by the remarkable presence of felspar and quartz, and in the disseminated crystals it contains, is considerably removed from the compact limestone that is seen beneath the serpentines in the parts of the Apen-

* It has been stated in a former note that the Bristol rocks do not belong to the submedial or transition class. [Translator.]

nines I have described. In the last formation, on the contrary, the general mass of the rocks is compact, sedimental (sedimenteuse), even arenaceous; and, whoever has seen both, will have perceived that the real and apparent differences are immense. M. Brochant says that the transition rocks of the Tarentaise ought to be regarded as the most ancient of that class; we assert that if the calcareo-sandstone of the Apennines is referred to the transition formation, it ought to be placed among the most recent of that class.

These subjects of comparison appear to me sufficient to establish the difference between the calcareo-sandstone rocks, (which, in the part of the Apennines I have described, are beneath the diallage rocks), and the ancient transition rocks, and consequently those only which may be regarded as belonging to a very distinct epoch.

If the same rocks are now compared with those I have mentioned under the name of inferior sediment rocks, and which are commonly called secondary Alpine rocks, we shall find many points of resemblance notwithstanding the differences still presented. I shall confine myself to giving as examples the rocks that I have had occasion to visit, and which, as much by the exact comparison which I have been enabled to make, as according to the general opinion of the most distinguished geologists, may with certainty be referred to the true Alpine formation.

§ III. *Comparison of the rocks beneath the Serpentines with the inferior sediment rocks termed Alpine.*

I shall take my first example on the southern side of the Alps, and on the shore of the lake of Como, from the town of Como to Nobialla, towards the middle of the lake and even a little beyond.

The rocks that border this lake, especially on the western bank, are, towards the base of the hills, a more or less bituminous blackish limestone in numerous beds, generally of small size, sometimes even thin enough to serve as the covering of a house instead of slates, alternating with blackish calcareous schist, and traversed by spathose carbonate of lime perpendicular to the fissures of stratification. These beds, though inclined in all directions, though contorted in every kind of manner, indicate a general rising towards the N.E., i. e. towards the primordial mountains, found at Bellano and Rezzonico.

Here there is a rock which, according to many geologists, presents very many of the characters attributed to the transition rocks; and, if it is added, that near the village named la Cadenabbia, I have observed in it sulphuret of zinc and madrepores, as in the limestone of Namur, Bristol, &c. I shall have nearly completed the characters of transition limestone.

But if, on the other hand, I add that this same rock contains a great number of fossil shells, such as ammonites, and especially turbines and bivalves resembling isocardiæ, all shells in too bad preservation to be determinable, that neither entrochi, nor orthoceratites are seen in it, many geologists would no longer admit it among these ancient rocks, which, according to them, do not contain any of the organic bodies I have mentioned.

Now, if even these rocks, which offer the characters of the transition formation much more decidedly than the sandy limestones of the Apennines, cannot with certainty be referred to it, ought not the latter to be attached to a still more recent epoch?

I should have a much greater number of examples on the northern end north-west side of the Alps. I shall content myself with mentioning three:

1. The Gemmi above the baths of Leuk, in the Valais, and that portion of the Alps that extends from this mountain to that of Pillon, or to the commencement of the Val d'Ormond, and which comprises the origin of the transverse valleys of Kander, Adelboden, Anderlenk, and Gsteig.

These mountains present, like those on the banks of the lake of Como, thin beds, extremely numerous, highly inclined, often sinuous, even contorted and twisted in every direction, but also often nearly horizontal; the rocks composing them vary but little: the principal or predominant are:

1st. A fissile compact limestone, of a brown, and almost black colour, passing into calcareous schist, and traversed by veins of spathose carbonate of lime mixed with quartz.

2dly. A shining calcareous schist, black, and as it were plastered with brown or greyish anthracite, and passing into spangled argillaceous schist.

3dly. A compact blackish limestone, containing greyish siliceous or sandy portions, sometimes in nodules disposed in the same line, sometimes in zones (descending into the valley of Wender-Eck), and passing into calcareous sandstone, even into quartzite (about Frutigen).

4thly. Spangled and blackish marly argillaceous schists.

I do not speak of the gypsum which occurs interposed, which is first seen at the Möserberg, and which, according to M. de Charpentier, continues to Bex and its environs. This circumstance, that does not occur every where, does not detract from the resemblance of the rest of the formation with that of the Apennines, and the examination I should make of it, would lead me too far from my subject. It will be remarked that all here is sandy and micaceous limestone, as in the Apennines, that the colours are there not deep, but that the compact rocks of an earthy and arenaceous appearance that compose its numerous strata, separate them so much from the other transition rocks mentioned above, that the greater number of geologists have considered them as of a more recent formation, or at least very different, and have assigned them the name of Alpine limestone, or formation.

If to these characters I add that fossil shells are found in it, though very rarely; that I have nevertheless found the impression of an ammonite or nautilus, geologists who will not admit this shell in the transition rocks,* will find in this fact another argument for separating the Alpine rocks of Oberland from the true transition rocks.

II. The mountain of Fis, to the N.E. of Servoz, in the valley of Sallanche.

It presents numerous and nearly horizontal strata, and the following rocks may be noticed as predominant and characteristic:

1st. Schistose rocks, very numerous, very little varied, mixed with mica, having the shining aspect of shining primordial schist, but differing essentially from it by the great quantity of carbonate of lime they contain;

2dly. A compact fine limestone, of a smoke-grey colour, with a scaly fracture, absolutely resembling that of Rochetta, and traversed like it by veins of spathose carbonate of lime.

3dly. Black calcareous schists, or dull argillaceous and non-calcareous schists, which resemble those of the banks of the lake of Como and the northern side of the Gemmi, and which contain like them ammonites, rare, certainly, but which appear to be all of the same species, as far at least as their state of preservation permits us to judge.†

III. The third example that I might bring forward, will be taken from the calcareous mountains of the environs of Glaris, from the valley of the Linth to Pantenbruck, and even Mont Dœdi. Not only do these mountains present the same brownish and blackish limestone, the same calcareous schist, the same spangled argillaceous slates as those I have remarked in the preceding Alpine mountains; but these rocks, which are blacker, more solid, and more sublamellar, possess still more of the characters attributed to transition

* It cannot however be denied, since M. Brochant has remarked, described, and figured a shell of this kind found in the marble named breche tarentaise, which forms part of the best characterised transition rocks.

† Care must be taken not to refer the fossil shells found on the northern side of the summit of the Montagne des Fis to this rock. They belong, as I shall perhaps have occasion to state elsewhere, to a formation altogether different.

rocks, yet they are still more than them considered as belonging to the Alpine or inferior sediment formation; and we there find, as in the preceding mountains, in the midst of a dull marly argillaceous slate, which appears almost homogeneous, which is black like the slate of Glaris, which, far from resting on the beds, forms part of it; we there find ammonites of the same species of that of the lake of Como.

The specimen I possess comes certainly from the calcareous schist of Oberhasli, a canton geographically different from that of Glaris, but geologically the same: for that country forms part of the calcareous chain which contains the Eigerhorn, and Mont Dœdi; the rock moreover that envelopes this ammonite entirely resembles the numerous marly argillaceous slates, or black calcareous slates of the mountains which have served me as examples; but as I shall probably have occasion to return to these mountains, in a note on the position of the fossil fish of this valley, I consider it sufficient to notice this part of the Alps in the number of rocks that may be compared with the calcareo-sandstone rocks of the Apennines. The calcareo-sandstone formations, essentially composed of compact smoke grey limestone, with spathose veins, of schistose marly limestone, of calcareous and micaceous sandstone, which, in part of the Apennines, are situated immediately beneath the serpentine formation, appear to me very different by their mineralogical characters, and their epoch of formation, from the *ancient* transition rocks, generally admitted as such, and which I have cited at the commencement of this comparison.

They even appear to me to possess more the characters of sediment, and to indicate a formation still more recent than the Alpine rocks I have just mentioned, and which are composed of brownish limestone, and micaceous argillaceous slates, sometimes containing organic remains, and generally described under the name of Alpine or inferior sediment rocks.

Article IV.

Opinions of Geologists on the position of the Serpentines and Diallage rocks.

I have said that the greater part, and perhaps even all geologists have entertained, and even published, on the epoch of the occurrence of serpentine, an opinion which appears to refer this rock to an epoch of formation much more ancient than the observations I have made on the Apennines appear to attribute to it. I shall not undertake to bring forward their opinions, that would be to repeat what is contained in the greater part of geological rocks; yet I cannot entirely pass them in silence, because some light may result from an examination of these opinions, which may lead us to a more precise distinction of the different serpentine rocks.

It is known that the geologists of Werner's school distinguish, with this father of geognosy, two formations of serpentine; the one, according to them, belonging to the primitive class, contains noble serpentine, granular ophicalce (ophicalce grenu), steachist, &c. and alternates with crystalline limestone, &c.; the other, which they refer to the last members or rocks of the primitive series, contains common serpentine.

All agree that the distinction of these two formations is difficult to establish, and consequently to recognise clearly. Now, by refering the serpentine rocks of the Apennines, which I have described, to the second formation, they are seen to be in a position that attaches them to an epoch much more recent than that of the *last primitive rocks*, or argillaceous transition slates. The opinion of the Italian geologists and those who have spoken of the geology of the Apennines, ought first and principally to occupy me. Messrs Viviani and Cordier have visited the environs of Rochetta, the mountain of Montenero to the E. of that village, and the situations of the manganese, the umber earth, the jasper, and diallage rock which it contains.

M. Viviani,* in 1807, when describing the mountain of Dragnon and that of Montenero, gives very exact details on the structure and nature of the diallage serpentine, the calcareous diallage rocks, and the jasper composing it; but he does not speak of their relative position, and regards diallage rock as primitive. This opinion was then so deeply rooted, that having remarked a serpentine crust which in some places covered the argillaceous ground of the southern slope of Montenero, he supposes that this mountain was formerly, and on this side, in contact with that of serpentine (p. 16), and M. Brocchi is disposed to admit this explanation.

M. Cordier, who visited this same mountain in 1809, and who has given the mineralogical and statistical account of the department of the Apennines,† having more attached himself to the technical part than geological considerations, has described the Montenero, the serpentines, the diallage rocks (under the name of granite de diallage), the jaspers, the manganese and brown ochre they contain, without explicitly speaking of the geological relations of these rocks with the sandy limestone of the Apennines.

He refers the diallage rock and all its varieties, the common schistose and diallagic serpentine, and the steaschist to the primary class; and the secondary class, the jaspers, the limestones, the marly schists, and blackish and calcareous slates worked at Lavagna, which afford a new proof of the alternation of the micaceous calcareo-sandstones with calcareo-argillaceous slates, marly schists, and limestones; he also refers to it the grey and blackish compact limestones, with spathose veins, and even the marble named *portor*, of Porto-Venere.

It was especially necessary for me to study the works, observations, and opinions of M. Brocchi, the geologist, who has, of late, given the best exposition of the structure of the Apennines. I fully appreciate the advantage I have had of conversing with him on the subject of this memoir, and I do

* Travels on the Apennines of Liguria, Genoa, 1807.

† Journal des Mines, No. 176, August, 1811.

not quote him solely for the purpose of compliment, but as an authority on which I depend, to give more force to the general conclusions I draw from my observations.

M. Brocchi * in the first place establishes, that the rocks which I have described under the names of schistose micaceous sandstone, and spangled argillaceous slate (phyllade pailletté), and which he names *macigno*, are not argillaceous schists; he refers them to greywacke, and he admits vegetable remains in it, at many points, as at Fiesole, Sestola, &c. but never marine shells, at least in that of Tuscany.

He refers this rock, the smoke grey compact limestone with a scaly fracture, &c. and the calcareous sandstone of the environs of Florence, named *pietra-forte*, to the transition class, yet notices an ammonite in it. He does not give these limestones the name of Apennine limestone; but he applies this name to a homogenous white limestone, containing chert (des silex), often blending its characters with the Jura limestone, but which differs from it by the fineness of its grain, &c.

He considers the serpentine as the primitive rock most generally spread over the Apennines of Eastern Liguria, where, he says, it is covered by transition limestone, argillaceous schist, greywacke, &c. and he cites Spezia, Monte-Cerboli,† and even the black transition limestone at Pian-del-Monte. It is accompanied by jasper at Fiegline near Prato (it is that of Monte Ferrato described above); he conjectures that the jasper forms part of the general ground, posterior to the serpentine, and even formed long after it.‡ He states the opinion of M. Bardi on the position of the jasper beneath the serpentine of Monte-Ferrato; but seduced by the generally received idea, that the serpentine is of primitive formation, he desires him to assure himself of

* Conchiologia fossile subapennina con osservazioni geologiche sugli Apennini e sul suolo adjacente. Milano, 1814.

† Tome I. p. 36.

‡ Ibid. p. 49.

the fact; * he says he has seen, in the same place, the jasper resting on the limestone, &c.

M. Cortesi † conceives that the serpentine of the Apennines belongs to the transition series, and considers it, as do all geologists, to be beneath the limestone of the Apennines, and on seeing it rise above this rock, he, with them, supposes that it pierces through the limestone formation with which it is covered at Gropallo, near the torrent of Nure, in the Parmesan states.‡

He places the sandstone (arenaria) beneath the limestone, considering it as the most ancient of the stratified rocks of the Apennines.

M. von Buch has said, in his travels in Norway (vol. 1. p. 476, French trans.) and in his memoir on Gabbro, that diallage rock is one of the most recent of the primitive class, and that this rock and serpentine are placed between primitive argillaceous schist and clay slate, and he cites Genoa, where, he says, the serpentine is beneath the argillaceous slate. He also cites Chiavari near Sesti, and Lavagna, the environs of Spezia, Prato near Florence, &c.

In the section of the Apennines from Bologna to Florence,

* This opinion is so general, that I am disposed to suspect that in many countries, even in Italy, there are two formations of serpentine rocks, not because in the environs of Genoa, as well on the north of that town as on the coast, the rocks with a serpentine base rest immediately on transition calcareous schists, which only proves that the calcareo-sandstone formation is wanting at that place; but because the greater number of Italian geologists admit them. M. Brocchi especially states in his memoir on the promontory of Argentaro and the Isle of Giglio: "that serpentine, very common throughout all the Siennois, occurs, as at the promontory of Argentaro, at the lowest part, i. e. beneath the argillaceous schist, the siliceo-calcareous breccia, the greywacke and limestone, that it consequently is the most ancient, and ought to belong to the primitive formation or most ancient transition period." At Falda-dello-Scalandrino, it is seen, he says, covered by the limestone that forms the summit of the mountain. (Bibliol: ital: 1818, tome xi. p 76, 237, & 356).

† Saggi geologici degli stati di Parma e Piacenza. 1 vol. 4to. Piacenza, 1819.

‡ Ibid. p. 91, & 127.

made by this geologist, and which was communicated in MS. to me, he has noticed serpentine in the environs of Pietramala and Corigliano; but he gives no information respecting its position.

I have stated, when speaking of the position of Monte-Ferrato, what M. Bardi has said, and I have cited Messrs Faujas, Viviani, Mojon, and Holland, when mentioning the serpentines of Monte-Ramazzo.

M. Marzari Pencati, in a notice he has published in the journal entitled, the Venetian Observer, for September and October, 1820, on the superposition and discordant mode of occurrence of granite on a secondary limestone, says a few words on serpentines. He notices a serpentine vein traversing the Alpine limestone at Canzocoli, and between Forno and Predazzo, in the valley of Avizio. He speaks of the passage of what he calls tertiary granite of three substances into serpentine rock. I feel much flattered at still finding myself of the same opinion with M. Marzari on this point. The diallage rock, named granitone by the Italian geologists, called diallage granite by M. Cordier, a perfectly crystalline rock, which possesses no volcanic character, which contains many of the elements of granites, occurring on a formation as recent as that I have described, disposes one to admit with less difficulty, the position of true granite on these same rocks. I have seen this granite at M. Mazari's, and it appears to me altogether like that of Cherbourg.

But there are two naturalists less read, even anterior to all those I have cited, who have perceived this fact.—The first is Ferber, who says, *that there are reasons for believing that the gabbro* (serpentine) *of the Imprunetta is placed on a limestone rock.. .. That it is a grey compact limestone, containing nodules of pyrites.* The state of science at the time in which he wrote (1772), did not allow him to draw any general conclusion from this observation.

The second is M. Palassou, that naturalist of Pau, who pursued to the end of a long life labours that evinced great activity, patience, and information. He had observed in the Pyrenees a rock, the characters, determination, and position of which had perpetually occupied him, and to which he has given the name of ophite. The ophite of M. Palassou is a greenstone (diabase), but it is an ill characterized greenstone, on one side passing into hornblende rock (amphibolite), in another into trappite (trappite), and in another into serpentine, as he observes, and as I have myself seen on the spot, at Pouzac, near Bagnères; this rock, which in every respect resembles that of Pietramala, rests like it on limestone, which M. Palassou refers to the secondary limestones. Here then at the foot of the Pyrenees, as at the summit of the Apennines, at a distance of more than two hundred leagues, is the same rock formation by crystallization, affording nearly the same modifications, situated in both places on a limestone of sediment. This fact surprizes us less now that it begins to become common; but it required, at the time M. Palassou remarked it, a good method of observation to see it, and some courage to declare it.

Geologists who have published general works have all thrown out the same opinion: it is found in Reuss, and all the English and German disciples of the Wernerian school. We shall confine ourselves to citing the most modern, those whose works have just appeared.

M. Breislak * considers, as do all the geologists he mentions, Messrs Cordier, Brocchi, Faujus, Viviani, &c. the rocks of serpentine as belonging to the last chains of primitive rocks, and neither cites in Italy nor elsewhere any examples of serpentine rocks of more modern formation.

M. Daubuisson, while admitting with geogolists of the Wernerian school two formations of serpentine, and referring the second to the transition epoch, does not find clear and authentic examples to give for establishing the epoch of formation of the last; he also regards diallage rock as belonging to the last term of the primitive series.†

M. de Bonnard, in his article Terrain of the Dictionnaire d'histoire naturelle, establishes, with all geologists, two formations of serpentine rocks; he refers the first, composed of diallage rock and serpentine often calcariferous, and in

* Institutions geologiques, 1818, t. I. § 276.

† Elemens de Geologie, 1820, t. ii. p. 160 & 170.

that case it is granular limestone, to the primordial rocks, and the second more compact, &c. to the last periods of that formation; he also cites the serpentines and diallage rocks of the Apennines, as dipping beneath the most ancient intermediate rocks. He premises that there exists much uncertainty and obscurity on the position of the two serpentine formations.

There apparently results from the facts and resemblances presented in this memoir:

1st. A tolerably exact knowledge of the relations of the serpentine and diallage rock with the jasper.

2dly. A precise determination of the rocks on which the preceding are immediately placed.

3dly. Numerous and authentic examples of the existence of serpentine and jasper rocks above a limestone of sediment, and above sandy and micaceous rocks of aggregation.

4thly. Direct proofs that the serpentine rocks of Rochetta, la Spezia, of Prato, of Pietramala, of Imprunetta, of the Volterranais, ought to be regarded as of posterior formamation to the rocks of sediment and aggregation, and strong presumptions, drawn from analogy, that the serpentine rocks of la Guardia, of Monte-Ramazzo, of the Bocchetta, &c. in the Apennines; of Musinet, Baldissero, and Castellamonte, at the foot of the Alps, and that even the rocks of serpentine greenstone (diabase ophiteux) of the Pyrénees, ought to be referred to this same formation.

5thly. Lastly, that rocks analogous to granites by their crystalline structure, again spreading over the surface of the globe after the existence of organic bodies, have covered rocks of sediment and aggregation containing the debris of these bodies.

This fact occurring with the same circumstances in very distant places, there is reason for believing that it has been as general as the most part of geological phenomena relating to the regular succession and sensible parallelism of the beds of the globe.

EXPLANATION OF PLATES V & VI.

Plate V Fig. 1 and 2.

Natural section of the two sides of the valley of Cravignola, near Rochetta, 15 kilometres [above 11 miles] to the N. of Borghetto, to the north of the gulf of Spezia.

Fig. 1. Right bank of the torrent.

Fig. 2. Left bank of the torrent, or base of Montenero, an escarpment opposite that of the right bank.

A. Green diallage serpentine.

B. Diallage rocks.

a. Serpentine diallage rock.

b. Calcariferous reddish diallage rock.

C. Reddish jasper mixed with greenish zones.

D. Smoke grey fine compact limestone, with spathose limestone veins.

E. Greyish marly schist.

Fig. 3. Section of a part of Monte Ferrato, near Prato di Sesto, to the N.W. of Florence.

A. Diallage serpentine.

B. Diallage serpentine passing into altered diallage rock.

C. Reddish and yellowish jasper.

D. Smoke grey compact fine limestone, in scattered blocks.

Fig. 4. Figure shewing the escarpment which is seen to the W. of the road after Pietramala, going from Bologna to Florence, between Maschere and Covigliano.

A. Hornblende serpentine.

B. Hornblende diallage rock and amygdaloidal diallage rock (euphotide variolitique).

C. Reddish jasper mixed with greenish zones.

D. Smoke grey fine compact limestone, with spathose veins.

E. Hard compact sandstone, and schistose sandstone.

Plate VI.

Fig. 1. 2. 3. Disposition of the serpentines, compact limestone, and dull micaceous argillaceous slate, observed at different points going from Poggio del Gabbro at the S. of Volterra, at the Lagonis of Monte Cerboli.

These sections are theoretical, i. e. that they are not the exact representation of the rocks observed, but only the graphic indication of the disposition of the rocks, such as it appears ought to result from the points observed at different distances, brought nearer in the figure, and placed in the relations which it is thought may be recognised between these different rocks.

Fig. 1. Hill named Poggio del Gabbro.

C. Fragments of jasper on the northern side of the hill.

D & E. Smoke grey compact limestone and marly schist, dipping beneath the serpentine A.

G. Gypsum which apparenly rests on the serpentine.

H. Clay alternating with the gypsum.

P. Siliceous conglomerate in an unconformable position.

Fig. 2. Descent towards the village of Monte-Cerboli.

B. Ill determined serpentine, mixed with diallage rock.

D & E. Smoke grey compact limestone, dipping beneath the rocks.

Fig. 3. Graphic indication of the disposition of the Lagonis, and the aqueous vapours in the valley at the foot of the hill of Monte-Cerboli.

D. Smoke grey compact limestone, &c. in highly inclined and much broken beds, in the southern portion, leading to Castel-Nuovo.

d. Indication of the debris and fallen calcareous masses towards the valley of the Lagonis.

M. A mass without structure of soft clay, of marly schist, of calcareous blocks and fragments, composing the bottom of the depression in the form of a basin, from which aqueous and sulphureous vapours rise up with violence.

Plate VI.

L. Lagonis properly so called, or small lakes of muddy and hot water, that traverse with impetuosity and violence the aqueous and sulphureous vapours containing boracic acid.

Fig. 4. Disposition of magnesite in the natural section of the hill of Castellamonte, near Turin.

A. Diluvium or alluvium (Terrain de transport) composed of rolled pebbles in its upper part, and of reddish sand in lower portion.

B. Mass of pale green disintegrated serpentine, in which veins of magnesite wind and inosculate.

a. a. a. Veins of magnesite.

b. Chalcedony in mammilated plates, &c. in the midst of some of these veins.

c. Veins of green concretioned hornstone.

d. Nodules or blocks of brownish green felspathic serpentine, scarcely altered in the centre, and decomposing in concentric layers.

On Fossil Vogetables traversing the beds of the Coal Measures: by Alex. Brongniart, *Member of the Royal Academy of Sciences, &c.*

April, 1821.

(Annales des Mines, for 1821.)

THE presence of organic remains in the midst of the solid and deep beds of the crust of the globe is, in the natural history of the earth, one of the circumstances most worthy of stimulating the curiosity and attracting the attention of observers.

These remains of ancient worlds, often so numerous and so little altered in their form or structure, though entirely changed in nature, seem to have been so well preserved solely in order to furnish us with the only documents we could ever obtain on the natural history of these different periods: they are as it were scattered phrases of that history. The more we collect them together, the more we may hope to establish, if not entirely, at least in its principal parts. The fact I am about to bring forward is not new; but the examples of this fact are still rare. It is besides so remarkable, so important for the theory of the formation of one of the rocks most interesting in every point of view, that too many examples cannot be collected.

That which is the subject of this notice is one of the most complete, the clearest and easiest to prove; it will therefore be one of the most authentic. I shall in this publication have no other merit, than that of having described and

G. Scharf Lithog:

COAL MINE OF TREUIL NEAR S^T

Shewing at F the compact Carbonate of Iron accompanying the Coal H.

Published by

PL: VII

Printed by C. Hullmandel

ETIENNE DEPART[T] DE LA LOIRE.

and at P, the Stems of large Vegetables in their Vertical Position.

W. Phillips. 1823.

figured, and consequently of having inscribed on the registers of science, a fact which Messrs Beaunier and de Gallois, Engineers of the mines of the department of the Loire, have shewn me.

It was long since known that coal deposites are accompanied by a great quantity of vegetable remains; it has also long since been remarked that vegetables resembling our ferns, and stems which do not exactly resemble that of any known plant, are most abundant in this formation; but it is not long since that it has been remarked that the entire system of these vegetable remains is different from the entire system of the same kingdom found in the more recent beds of the globe; lastly, but few years have elapsed since it has been recognised that these vegetable remains were not always extended between the fissures or on the surface of the beds, and parallel to their stratification, but that they in some places cut them, that they traversed them in many, that they were even perpendicular to them, and lastly, that they sometimes occurred in the vertical position in which vegetables usually grow.

If these ideas had been more generally spread, if the facts which they establish had not been considered as exceptions owing to chance, theories would not have been proposed, even lately, on the formation of coal, which are in evident contradiction with these facts.

The vertical stems we are about to describe, have already been mentioned by M. de Gallois; they occur in the most distinct manner at the mine named du Treuil, at a 1000 metres [3076 feet] to the N. of St. Etienne, department of the Loire.

The coal formation offers in this place two circumstances, which are rare, but very favourable for observation; it occurs in beds evidently horizontal, and situated in such a manner, that it can be worked in open day, and as a quarry, so that it has furnished us with a very uncommon opportunity in this kind of formation, of observing a natural and complete section of the different rocks and minerals composing it, and of being able to represent them with a clearnesss and extent that subterranean workings can never offer.

This natural section of the formation is not only interesting from the circumstance of the fossil vegetables that form the principal object of this notice; but also for the presence of compact carbonate of iron ore, which so constantly accompanies the coal, and which will soon become in France, as it has long since been in England, the object of great research, and a species of industry new to us.

By confining the examination in the Treuil mine to the portion represented in the view (pl. 7), that is joined to this notice, there is remarked passing from the bottom to the top, i. e. from the lowest terrace to the surface of the ground;

1st. A bed of coally spangled clay-slate (*s*), which is soon followed by a bed of coal (H), about 15 decimetres [5 feet] thick.

2dly. A second bed of the same slate (*s*), but thicker, and containing in its lower strata and very near the coal, four beds of compact carbonate of iron ore (F), in flattened nodules, cleanly separated from each other, more or less voluminous, or in large plates swollen in the middle, accompanied, covered, and even penetrated by vegetable remains;

3dly. And at the second terrace above this schistose bed, another bed of coal (H), which is from 46 to 50 centimetres [about 20 inches] thick, and which is covered by a bed composed of schistose clay (*s*), resembling the inferior one, from four to five thin seams of coal, and towards its upper part of three or four thinner seams, and closer together, of carbonate of iron (F), in every respect resembling that of the first terrace.

The schists and iron ore are accompanied by numerous vegetable impressions, which cover their surfaces and follow all their contours;

4thly. Lastly, and here terminating the coal measures, a bed 3 or 4 metres [about 9 f. 10 in. to 13 f. 1 in.] thick occurs, of a micacous sandstone, sometimes simply split in various directions, sometimes very clearly stratified, and even passing in the mass into the slaty structure.

In this bed, and over a great extent, numerous stems occur, placed vertically, traversing all the strata, and of which the view joined to this notice only shews a small number. It is a true fossil forest, of monocotyledonous vegetables, resembling bamboo, or a large *equisetum* as it were petrified in place.

Although the beds of the coal measures are here evidently horizontal, it is remarked that there has been, after the precipitation and even consolidation of the upper sandstone, a sliding movement, of little extent certainly, but sufficient in many points to break the continuity of these stems; so that the upper parts are as it were thrown on one side, and are not continuous with the lower.

It does not enter into my plan to describe these vegetables, nor to search to determine to what family they may belong: it is a very important and difficult subject, and cannot be treated en passant. My son, aided by the counsels of M. Decandolle, and the help of geologists, has long since undertaken a special work on this portion of botany, the object of which is, the study of fossil vegetables; for by too superficially and rapidly naming the vegetables of the coal measures, one risks propagating opinious on their genera, which might be erroneous. But although I ought to speak here only of the position of these stems, and not of their nature, I cannot avoid offering, under this point of view, some observations relating directly to those of St. Etienne.

There are at the Treuil mine two very distinct kinds of stems: those of one class are cylindrical, articulated and striated parallel to their edges; they do not in their interior present any organic texture, their *probably* reedy (fistulaire) cavity is entirely filled with a rock of the same nature as that composing the beds they traverse. These stems are the most numerous, they vary much in diameter from 2 or 3 centimetres only [about 1 inch] to 1 or 2 decimetres [4 to 8 in.], and perhaps beyond. Their greatest length appeared to us to be 3 or 4 metres [9 f. 10 in. to 3 f. 1 in.]. Their surface is often covered by a ferruginous and even coaly deposit or coat.

The other and more rare vegetables, are composed of hollow cylindrical stems diverging towards the lower extremity, and appearing to spread out in the manner of a root, but without presenting *any ramification.**

None of these stems can apparently be referred to trees of the palm family. This result, which I only notice, will be developed and preceded by the motives which have led to it, in the special work my son will publish on this subject.

I announced at the commencement of this notice, that the fact described in it was not new to geologists. Among the examples brought forward of fossil vegetable stems traversing many beds, or placed vertically in the bosom of the earth, I shall speak of those which appear to me most analogous to the example taken from the St. Etienne mines: these citations will help to establish the resemblances that are as real as remarkable, which the coal measures of all countries present, under all the circumstances of their formation and structure.

Sir G. Mackenzie has observed in the coal measures of Scotland, near Pennycuick, ten miles from Edinburgh, a vertical trunk about 12 decimetres [4 feet] high, the mass of which is of coal measure sandstone, and of which the bark or what, here represents it, is replaced by coal. This trunk not only appears striated in the manner of the St. Etienne stems, but also divided like them by transverse articulations.†

A fact closely resembling it appears to have occurred in the coal measures of South Shields.‡

M. de Schlotheim also mentions vertical stems at Kiffhausen, in the Hartz,‖ in the mines of Manebach, near Ilmenau, &c.

But the examples that most approach that which I have brought forward, have been observed in Saxony by Werner,

* The figure shews these different circumstances.

† Biblioth: universelle, t. viii. p. 256.

‡ Ibid. t. viii. p. 234.

‖ In Leonhard Taschenbuch 1813, 7th year, p. 40.

by Messrs Voight and d'Aubuisson, in the coal measures in the neighbourhood of Hainchen, and by Messrs Habel and Noggerath, in the coal mines of the Saarbruck country.

Four or five stems, from 20 to 30 centimetres [8 to 12 feet] in diameter, which M. d'Aubuisson calls trunks of trees, occur at the first spot in a vertical position, in the sandstone of the coal measures. All the circumstances agree with those that accompany the vertical stems of St. Etienne.*

The same facts have been observed in the environs of Saarbruck, in many coal mines, especially in that of Kohlwald, where the trunks are 2 metres [about 6 feet 7 in.] high, and 6 or 8 decimetres [2 feet to 2 feet 8 in.] in diameter, and in that of Wellesweiller; the trunks in this last mine, remarkable for their conical form, for their diameter of from 45 to 36 centimetres [18 to about 14½ inches] for their height, which is above 3 metres [about 10 feet] have lately been described and figured by Dr. Noggerath.†

These trunks, which cannot be referred to any known vegetable, and which appear to differ from those of Hainchen and St. Etienne, traversed many beds of sandstone, both schistose and sandy, and were situated between two coal beds.

M. de Charpentier cites a similar fact, which he had observed in the sandstone of the coal measures to the N.E. of Waldenbourg, in Lower Silesia. He states that, in 1807, he there found a fossil tree in a vertical position, traversing horizontal beds, and having its roots and some branches well preserved, and changed into very small grained quartz, of a greyish black colour, but whose structure was no longer discernible; the bark and small branches were changed into coal. This trunk was 4 decimetres [16 inches] in diameter, and there still remained a length of about 4

* See Journal des Mines, t. xxvii. p. 43, and especially d'Aubuisson, Geognosie, t. ii. p. 292.

† Ueber aufrecht in gebergsgestein ingeschloffene fossile Baumstamme, &c. von Dr. Jacob Noggerath, Bonn, 1819.

metres* [13 f. 1 in.]. The presence of branches, which does not appear doubtful, establishes a remarkable difference between this fact, that of St. Etienne, and those we have quoted.

Lastly, M. Habel has observed in these same mines vegetable stems placed almost vertically, which do not differ from ours; they were from 2 to 2½ metres [about 6 f. 7 in. to 8 f.] in height, and about 25 centimetres [10 feet] in diameter: they were articulated, regularly grooved, and covered by a little coal. These stems traversed beds of the formation which contain carbonate of iron ore.

A trunk of a tree in a vertical position has lately been observed in the sandstones which cover the coal formation of Glasgow, on the N.W. of that town: this trunk was about 6 decimetres [2 feet] in diameter, its transverse section presented a nearly oval figure; it was, like those I have described, entirely filled with the rock composing the matrix in which it occurred; but the bark, that is to say, the exterior part of this vegetable, for nothing shews that it had a real bark, was converted into coal. It was cleared for the extent of about a metre [about 3 f. 3⅓ in.] and no branches were observed; yet at its lower part, roots were said to have been seen, especially four of large size, dipping under the ground like those ordinary trees. It could not, says the author of this notice, be referred to any known tree. (Thomson, Annals of Philosophy, 1820, November, p. 138).

I do not speak of the stems and trunks of true trees, not only fossil, but changed into silex, which are observed in the rocks of a formation absolutely different and always posterior to that of the coal; these petrified woods are very numerous, but their geological position distinguishes them essentially from those which form the subject of this notice.

It is probable that the examples of stems traversing the beds of the coal measures are also very frequent, and that if but a small number have been noticed, and so few figures of them have been published, it is owing to the manner in

* Biblioth: univers: 1818, t. ix. p. 256.

which the rocks containing them are reached. These rocks are almost always deep; they are arrived at only by pits and galleries, which are never much developed in many directions. When forming the subterranean passages, the sandstones are avoided as much as possible, as they only offer to the miner expense without profit; yet these are the rocks which appear to contain most of the vertical stems. The difficulty of uniting all these conditions, ought to diminish the number of circumstances favourable to the discovery and easy and complete observation of these stems; but analogy leads us to believe that, if there was the same interested motive in searching for them as the iron ore, they would be found as generally spread over the coal measures as this ore. Now, if these stems, still in their vertical position, shew that the coal measures of St. Etienne, Saarbruck, &c. have been formed and deposited on the spots where these vegetables have lived, as much might and ought even from analogy be said of all other coal measures. The arborescent ferns and all the vegetables of a tropical aspect found buried in the coal measures must no longer then be sought to be found beneath the torrid zone, nor to be brought into our latitudes by means of great currents or grand debacles. This hypothesis, now almost entirely abandoned, is, as M. Noggerath has particularly remarked, incompatible with a vertical and general disposition, which is so clear and so general.

Yet M. de Charpentier, in the notice we have mentioned, and which relates to the vertical trunk of Waldenburg, offers very just reflections on the difficulty of conceiving that these stems could grow in a rock such as that which now envelopes them, and that this rock could deposit itself in the middle of them during their growth, without in part destroying, upsetting, or at least deranging them. He supposes that these vegetables, adhering by deep roots to the ground, have been carried away with the soil that supported them, and left in the places where they are now observed. He rests this explanation on a fact which he observed at the time of the great debacle of the lake of Bagne. In this great catastrophe, large trees with their roots were carried away by this debacle, and deposited vertically in the plain of Martigny. This observation leads to the conclusion that the vertical position of a stem is not a proof that it has lived on the spot where it is now thus found; but it appears to us a circumstance that ought to be rare, and which can only offer some isolated facts: the examples of vertical stems are on the contrary very numerous. In those mentioned by M. Noggerath and us, it is not only one large trunk that has been observed, but many; and in that of the Treuil mine, which forms the principal subject of this notice, there is nearly a forest of slender stems, which have preserved their parallelism among themselves. Moreover the nature of the ground to which the vegetables still held by their roots ought to be different, or at least very distinct from that of the rock enveloping them. It is perhaps more difficult to conceive that this sandy rock could envelope them after their removal without deranging them, than that it has been deposited between them in the place where they grew, and where they were very firmly rooted. Even supposing that these vegetables could have been transplanted without losing their vertical position, it cannot be admitted that they came from a great distance; and the insurmountable difficulty that this fact raises against the hypothesis which brings the coal plants from tropical regions into our climates, does not the less remain.*

Yet the reflections of M. Charpentier and the facts he cites, throws uncertainty on the primitive situation of these vertical stems, and ought to engage us to continue our observations, and teach us that we cannot draw any absolute and general conclusion from this fact.

* For an account of the vegetables of the coal measures, consult Conybeare and Phillips's Outline of the Geology of England and Wales. Part i. p. 333 to 343. (Trans.)

Notice on the Coal Mines of the Basin of the Aveyron: by M. le Chevalier Du Bosc, Engineer of the Royal Mining Corps.*

(Annales des Mines, 1821.)

THE department of the Aveyron is without contradiction, of all those in the kingdom, that in which the thickest as well as the thinnest known beds of coal are worked. This assertion at first sight strange, ceases to be so when the different parts of this department are visited, in which the coal mines occur. Enormous beds of coal will be seen in the Canton of Aubin, some of which are more than from 15 to 20 metres [about 49 feet 6 in. to 66 feet 6 in.] thick, and in the Cantons of Milhau and its vicinity, coal seams are worked which are generally not more than from 20 to 25 centimetres [8 to 10 inches] thick.†

[The author then proceeds to state that there are three distinct coal basins in this department, including that of the Aveyron—viz.]

1st. The coal basin of the N.W or of the Lot, which belongs to the sandstone formation, ‡ and contains the rich mines of Aubin.

2dly. The centre coal basin, or of the Aveyron, which belongs to the sandstone formation ‖.

* In the coal basin of the Glane, and the rivers flowing into it (Palatinat) numerous coal mines are opened on beds which are less than 2 decimetres [8 inches] thick.

† This is extracted from a detailed account of the mode of working the mines, &c. in this district. (Trans.)

‡ and ‖ Coal measure sandstone I presume. (Trans.)

3dly. The coal basin of the South, or of the Tarn, which belongs to a limestone formation.

Of these three coal basins I shall only here speak of the second, that of the Aveyron.

Its general direction is from E. to W. and it runs nearly parallel with the river Aveyron, constantly following the left bank of this river, without ever passing over to the right side, excepting on its western limit, (very near the town of Rodez).

The two towns of Rodez and Severac-le-Chateau may be considered as the two extreme points of the coal formation, which is thus about 36 kilometres [about 27$\frac{1}{3}$ miles] long from E. to W. while its breadth from N. to S. is variable, never exceeding 3 kilometres [about 2$\frac{1}{4}$ miles], and being most commonly much less.

The coal formation, beginning very near Rodez, the chief town of the department, passes along the left bank of the Aveyron, traversing successively the communes of Agen, Laloubiere, Montrozier, Bertholene, Layssac, Severac-l'Eglise, Gaillac, Recoules, and Laverahe. Coal has been found and worked in these different communes, with the exception of those of Severac l'Eglise and Galliac, where there has been found, not coal, but coal measure sandstone.

The limits of this coal formation are, on the N. a vast formation of secondary limestone which covers the sandstone, and which forms a vast platform, named *Causse*, * which occurs between the rivers Aveyron and Lot. The predominant rock is a true compact limestone, almost always shelly, but very variable in its colour, structure, and hardness. Its colour varies from yellow to dull white and grey; it sometimes passes into argillaceous limestone; at others it becomes so schistose, that it is employed as roofing slate.

The direction of the limestone beds is from E. to W.; their inclination towards the S. is always slight, and it has

* The word *causse* is a generic name for the limestone soils proper for the cultivation of wheat, while the name *segala* is given to the mica slates and gneiss, which scarcely produce any thing but rye.

often none at all. Its inclination is not sensible except towards its point of contact with the coal measures; the beds there become inclined from their previous horizontal position, and acquire a dip of from 30° to 40°.

Towards Severac-le-Chateau some limestone strata contain veins of lignite (true jet) of little thickness, and not continuous enough to be worth working. The coal formation is bounded on the S. by primitive rocks, on which it rests. The rocks are composed of gneiss, the beds of which, not very distinct, incline to the N. Proceeding S. the gneiss is replaced by granite, which constitutes the central and elevated chain of Levezon, separating the basin of the Aveyron from that of the Tarn.

The predominant rock of the coal measures is the true coal measure sandstone, in beds of more or less thickness, more or less regular, but always inclined towards the N. or nearly so. This sandstone is sometimes large grained, but most commonly fine-grained. Though very hard to be worked, it speedily decomposes in the air, loses its hardness, and becomes of a yellowish colour on the surface. It sometimes alternates with micaceous sandstone, and sometimes retains impressions of ferns and reeds.

These impressions also occur in the argillo-bituminous schists, which, in various points of the formation, accompany the coal beds, either as a roof or floor, or dividing the coal-beds themselves.

Besides the schist, small veins or nodules of soft clay, sometimes schistose, are found on the sides or in the interior of the coal beds.

Lastly, the sandstone contains beds of coal, which, in different points, vary as to thickness, inclination, and quality. Their dip is the same as that of the sandstone.

The number of coal beds observed at different points varies from one place to another; it is the same with their mode of occurrence, thickness, and quality.

Notice on the Geology of the Western part of the Palatinate, by M. de BONNARD*, Engineer in Chief of the Royal Mining Corps.*

(Annales des Mines 1821.)

THE mountainous country on which I now propose to offer some geological remarks, collected fifteen years since, in several professional tours, comprises a part of the ancient departments of the Sarre and Mont Tonnerre: it is limited on the W. and N. W. by the course of the Brems and that of the Nahe; on the S. by the present frontier of France; on the E. by the prolongation of the Vosges chain, to the fort of Mont Tonnere; lastly on the N. E. by a curved line passing within the limits of the small towns or bourgs of Gælheim, Alzey, Wællstein, and Creutznach. Beyond the latter limit, and the red sandstones of the Vosges, are the rich plains of the eastern part of the Palatinate, the soil of which is formed of less ancient rocks than those of the western part.

The river Nahe flows in a general direction from S. W. to N. E. from its source, situated near Selhach at the foot of the mountain of Schaumberg, to the small town of Kyrn, where it turns towards the E. It follows this last direction to the *salines* of Creutznach; here it again turns, and flows to the N. and even the N. N. W. to its junction with the Rhine at Bingen, a point that occurs nearly in the prolongation of its original direction.

On the left bank of the Nahe, and at a short distance from its bed, the schistose and compact quartzite formations commence, which form the mountains of the Hünsdruck. On the right bank, and also at a short distance from the river, are situated the coal measures and red sandstones of the Palatinate.

The Hünsdruck, bounded by the Rhine, the Moselle, the Sarre, and the Nahe, forms part of the great schistose zone which is prolonged from the deportment of the Ardennes across the N. of Germany, and which appears in a great measure formed of transition or intermediate rocks. The red sandstones of the Palatinate join, on the E. those which constitute the mountains of the whole northern part of the Vosges, known in the country by the name of Hardt-geberge, the eastern slope of which is rapid, but which gradually declines to the west, towards the country which especially forms the subject of this notice. In this chain, the granite, long hid beneath the secondary rocks, appears for the last time between Landau and Annweiler; it there forms near Alberschweiler an isolated mountain, in which the granite rock is seen to pass into porphyry. This mountain rises in the midst of the red sandstone that surrounds it, and which immediately rests upon it. Proceeding from this place towards the N. to the foot of Mont-Tonnere, or fowards the west to Sarrebruck, red sandstones and quartzose conglomerates are only found, the whole of which is commonly known by the name of the red sandstone formation. They are covered, but only in a few points, in this direction, by horizontal shelly limestone (muschelkalk), as at Bischmissheim, near Sarrebruck, or by limestone and marly clay, as in the environs of Deux Ponts, or by gypsum placed between the red sand and limestone, as at Omersheim, between Sarrebruck and Bliescastel. Not far from Sarguemine, on the right bank of the Sarre, is situated the small *saline* of Relchingen, near the limit commou to the red sandstone of the Palatinate, and the horizontal limestone of Lorraine: the spring only contains about 1½ per cent. of salt, with muriate of soda, and sulphates of soda and lime. Still more west the red sandstones envelope the southern part of the coal measures, are prolonged on the left bank of the Sarre, to and beyond the environs of Treves, and even penetrate, on the right bank of this river, into the basin of the Brems and its confluents. They are also, in some points, covered with horizontal limestone, as at Nalbach (two leagues to the north of Sarre Louis,) at Wahlen (between Mergiz and Wadern,) &c. It is probable that this great mass of arenaceous rocks comprises the two formations of red sandstone, known in Germany by the names of Rothe liegende and Bunter-sandstein.* I have observed vegetable remains half carbonized in the sandy rock near Sarrebruck.

The coal measures form a zone which extends, from S. W. to N. E., 25 leagues in length, from the southern bank of the Sarre, a little below Sarrebruck, to beyond the Nahe in the environs of Sobernheim. The breadth of this zone of coal varies from four to seven leagues, according as it is more or less confined by the two chains between which it occurs. At about a third of its width it is traversed by a band of the red sandstone formation, which constitutes some elevated summits, among others that of Höcherberg, near Waldmohr, and which divides the coal measures into two basins very different from each other.

The southern basin which sheds its waters into the Sarre, belongs to the best characterized and richest coal measures. It is principally composed of alternating beds: 1st. of argillaceous schist, slaty clay, and schistose sandstone, in which are observed numerous impressions of ferns and other plants common to this formation; 2dly. of micaceous sandstone or coal measure sandstone; 3dly. of argillaceous and quartzose conglomerates. It contains good and numerous beds of coal worked in the environs of Sarrebruck, as also beds and abundant masses of earthy carbonate of iron ore, in the nodules of which are sometimes remarked impressions of fish, particularly in the upper part of the coal measures, as in the environs of Lebach. The schists of the coal measures are in some places worked in order to extract the alum and sulphate of magnesia: a small and slightly salt spring rises from it near Sultzbach. This formation also contains, but only between its upper strata, beds of compact limestone, grey or black, with a splintery fracture, and sometimes a schistose structure.

The general direction of the beds in this basin is S. W. and N. E. On the east and north, it would appear that this for-

* The new red or saliferous sandstone. (Trans.)

mation rests upon the red sandstone that surrounds it, and whose beds appear in some places, on the banks of the Blies, near Neunkirchen, to the S. E. of Ottweiler, to dip beneath the coal measures. On the S. and W., on the contrary, the coal measures dip beneath the red sandstone, and are found by traversing the sandstone.*

The northern part of our coal zone, which sheds its waters into Nahe, and which principally comprises the banks of the Glane and its confluents, is of a different nature. The argillaceous schists, with little or no impressions, often form the principal mass of this formation and sometimes entirely constitute it: they commonly alternate with schistose sandstone; but the variety of sandstone especially known by the name of the coal measure sandstone, is rather rare. A coal almost always dry and of bad quality often occurs in these rocks, forming in each mountain one or at most two small beds of a few inches thick, in general situated near the surface. The coal is nearly always immediately covered, and also sometimes divided into two beds, by a limestone of a dull yellow or blackish brown, or presenting different mixtures of these two colours, and sometimes containing patches (mouches) of sulphuret of zinc.

The coal and limestone are worked together in numerous small mines, and the principal use of the coal is to burn the lime, which has been generally employed as manure for forty years, and has singularly improved this barren country.

Marno-bituminous schists have been observed in the same formation, which sometimes present (at Munster Appel) the impressions of fish penetrated with sulphuret of mercury†;

* It would appear that part of the red sandstone of the Palatinate, that which supports the coal measures, might be referred to the old red sandstone, the carboniferous limestone being wanting, and that the part resting on the coal measures was the new red or saliferous sandstone. Some of the lower beds of the coal measures, or the mill-stone grit, (if present) may however be red. (Trans.)

† This very remarkable circumstance, which reminds us of the marno-bituminous schists of Hesse and the Mansfield country, the bad quality and slight thickness of the coal, the uniformity with which it is covered by a limestone very much resembling zechstein, &c. led me to consider

it also contains large grained quartzose conglomerates, the cement of which, of a reddish-brown colour, often appears derived from the destruction of trappean rocks. Lastly, the beds of compact limestone, of a dark colour, resembling those met with in the western part of the basin of the Sarre, between the upper strata of the coal measures, occur here in very frequent beds in the midst of the schists and conglomerates, and even in many places appear (near Wolfstein, Rothseelberg, &c.) beneath the whole coal formation.

No general direction can be observed in the stratification of the Glane coal basin. The most southern coal beds, which are the best of the whole basin (those of Altenkirchen and Dorrenbach) incline to the N. and thus appear to rest on the band of red sandstone separating them from the Sarre coal basin; but more on the N. the beds of coal worked often incline nearly parallel to the slope of the mountains that contain them, and the general disposition of the rocks appears to be determined by the inequalities in the surface of an inferior rock, situated at a slight depth.

It is in the coal formation of the Glane that a great part of the mercurial mines of the Palatinate is worked; the ore either forms veins, as at Mærsfield, Potzberg, near Cousel, &c. or more or less irregular masses, as at Stahlberg, and Landsberg, near Obersmoschel. Traces of lead ore have been noticed in the same formation, but which have not given rise to any works. Many slightly salt springs are known in it near Grumbach, at Diedelkopf, near Cousel, and elsewhere; it is said that the spring of Diedelkopf had formerly been worked.

On its N. E. limit, the coal formation is covered, in the environs of Alvey, by horizontal limestone, which extends on the N. and E. to the banks of the Rhine.

that the coal formation of the Glane was but the equivalent or representative, on the left bank of the Rhine, of the marno-bituminous schist formation, spread over the centre of Germany, and which sometimes also contains coal but on the other hand wide differences seem to oppose themselves to the adoption of this idea, whieh I was unable to submit to a severe examination, as I did not return into the Palatinate after I had observed the rocks in the Mansfield county.

The line of separation formed by the course of the Nahe, between the rocks of Hunsdruck and those of the Palatinate, is prolonged in the same direction to the S.W. and then nearly follows the basin of the Brems, which appears to form the prolongation or pendant of the basin of the Nahe, on the other side of the Schaumberg, which serves as the point of division of the waters. But in that part of its course where the Nahe turns out of its general direction (between Kyrn and Creutznach) it no longer corresponds with the limits of the two formations; I have not any where found, on the right bank, the schists and quartz rocks, except at the spot where it joins the Rhine; but, as I have already noticed, the coal measures and even the red sandstones, in many points penetrate on the left bank, especially on the north of Sobernheim, where traces of coal have been observed on the slope of the mountains of the Hundsruck,* and on the N.W. of Creutznach, where the red sandstone forms the surface of the country for some extent.

The course of a river between two formations of different natures and epochs, is a fact frequently observed in geological investigations, and which often appears to support the idea of valleys having been formed by running waters; it is then thought that the rocks, generally less compact in the neighbourhood of the limits common to two rocks, may have been more easily attacked and destroyed by the waters. But it will be very difficult to draw the same conclusion from the examination of the valley of the Nahe. In fact, between the two formations of the Hunsdruck and the Palatinate, there occurs a trap formation, composed of much harder rocks than those of the two formations it separates; and it is in these trap rocks that the bed of the Nahe is hollowed out for nearly its whole course, between two steep banks, a circumstance the more remarkable, as the breadth of the trap formation is often very inconsiderable, and that it is

* Similar traces are known, in the same position, near the forges of Abentheuer, on the W. of Birkenfeld; i. e. on the W. of the trap zone, of which we shall presently speak; this is an exception to the general disposition of that zone between the two formations of the Hunsdrück and Palatinate.

sometimes limited to the mountains which immediately confine the bed of the river. This trappean zone is also prolonged on the S.W. into the basin of the Brems, but in a less continuous manner than on the banks of the Nahe.

On the left bank of the latter river, the trap suddenly and completely disappears on arriving at the schistose formations: this limit offers very interesting geological points of view, from the different aspect presented by the mountains of the two formations. The soil of the Hunsdruck is generally of a greyish tint; vast and level platforms are there observed, with scattered blocks of white quartz, in general well cultivated, though not fertile, and covered with fine forests of beech and oak. These platforms are furrowed by valleys with steep sides, but uniform and generally covered with vegetable soil. On the trap formation, a reddish brown soil is seen, of nearly general barrenness, and scarcely wooded; the summits of the mountains present rounded hummocks, but their sides are broken and expose numerous escarpments of blackish rocks. Vast excavations cut on the surface for the extraction of the iron ores, almost exactly mark the limits of the two formations. On the right bank of the Nahe, on the contrary, the trap is found in the midst of the coal measures of the Palatinate, as far as eight or ten leagues from the river, forming either isolated hills, or branches of mountains less elevated than the principal chain, but generally directed like it from S.W. to N.E. This trap formation is principally composed of *corneans*,* wackes, greenstones, and amygdaloïds (the latter rock contains agates, chabasites, prehnites, &c. in the environs of Oberstein). Sometimes the cornean passes into flinty slate (kieselschiefer) well characterized, and of a black colour; but it also passes into a basalt, and all the rocks become entirely analogous to basaltic rocks. In some localities, particularly near St. Wendel, and in the valley of Oberwiesen, on the north of

* I have adopted the word "Cornean" as a translation of the French "Cornéanne." In a paper on Southern Pembrokeshire, lately sent to the Geological Society, I have had frequent occasion to mention this kind of rock; the reasons for adopting the word Cornean are there stated. (Trans.)

Mont Tonnerre, I observed among the trappean rocks a remakable mixture of talcose parts, and silky asbestos. Breccias and conglomerates are also observed in this formation, composed of fragments of cornean and quartz, sometimes very large,* at others so fine that the rock insensibly passes into a red sandstone (rothe liegende). These arenaceous rocks commonly appear above the crystalline rocks, often also the superposition of the red sandstone on the trap is very distinct, as is seen near Winnweiler; but sometimes also the whole appears disposed in alternating beds, and in this case an insensible passage may often be observed from the cornean or wacke to a fine grained sandstone: I have observed the latter fact at the Schaumberg, near Tholey, on the N. of St. Wendel, in the hills which border the Brems near Wadern, on the banks of the Nahe near Durckroth, and elsewhere. Sometimes the paste of the cornean, by alteration, becomes whitish, soft, and unctuous; it is then worked as fuller's earth.

From the environs of Birkenfeld to below Oberstein, the principal trap band is many leagues broad on the right bank of the Nahe, where it extends to beyond Baumholder. Throughout all this country, the coal measures rest on the slope of the trap mountains. Further to the N.E. the latter are more contracted; in the environs of Kyrn, the coal measures traverse the Nahe in many places, and the trap rocks occur above them, and it is then that they especially become analogous to the basalts. I have observed the superposition of these rocks on the coal measures, principally on the banks of the Nahe, near the coal mine of Durckroch, and near Hefersweiler, 2 leagues to the E. of Wolfstein. An analogous fact it is imagined may be observed in the mountains of Landsberg and Stahlberg, celebrated for their mines of mercury; but the whole interior of these mountains appears so much disturbed, that it is very difficult to determine any real superposition. On the contrary, I have seen

* A singular breccia formed of fragments of cornean cemented by a cornean paste or base, occurs associated with trap at Cuffern mountain near Roch, and in Ramsay Island, Pembrokeshire. (Trans.)

on the banks of the Lauter, near Olsbrücken, and in many other localities, the coal measures and coal resting on amygdaloid. Be that as it may, the preceding facts appear sufficient to prove that at least a part of the trap rocks of the Nahe should be considered as contemporaneous with the coal measures of the Palatinate; but it must also be recollected that M. Omalius d'Halloy, and M. Calme, have thought that they observed these rocks in some points, beneath the ancient schists of the Hunsdrück,* on the left bank of the Nahe.

The trap formation contains large veins of sulphate of barytes; one of this kind is seen near and to the N. of Baumholder, on the road to Oberstein. It is stated that near Seelen (to the E. of Wolfstein), a considerable vein of reddish spathose limestone is worked in this formation. It also contains such a great quantity of small veins or bundles of copper ore, that it is known in the country by the name of Kupfergebürge (copper formation). The copper mines of Fischbach, Norfeld, Baumholder, Oberstein, Niederhausen, &c. were worked in it; all are now abandoned. Traces of mercury have been observed in this formation at Baumholder, St. Julian, and elsewhere; the manganese vein of Crettnich near Wadern, is worked in it, as also many iron veins at the foot of Mont Tonnerre; lastly must be cited the great masses of iron ore which are in many places worked in open day, on the limits of the trap of the Nahe, and the schists of Hunsdruck.

But independantly of the trap formations, well character-

* See Journal des Mines, No. 144, and No. 146, p. 144.

In a work entitled Geognostische Studien am Mitttel-Rhein, printed at Mayence in 1819, M. Steinheimer cites, page 112, &c. numerous localities in the Palatinate, where basalt, wacke, and amydaloïd, apparently form beds in the coal measures. In other places, on the contrary, as at Braunhasen, on the slope of the Hunsdrück, he considers the basalt as belonging to the transition rocks of the schistose mountains (p. 200). From the whole of his observations he concludes, that all this trap formation is probably the product of submarine volcanoes, which acted on the pre-existing rocks of Hunsdrück, at the same time that the sea deposited the secondary formations of the Palatinate.

ized formations of petrosilicious porphyry, and argillaceous clay porphyry (argilophyre of M. Brongniart), have also been observed in the Palatinate. The predominant rock of this class has a whitish, greyish, or roseate paste, and contains crystals of felspar, quartz, mica, and sometimes hornblende.

These porphyries occur on both sides of the trap zone: on the western side I have observed them, proceeding from S. to N. 1st, at Dippenweiler (3 leagues on the N. of Sarre-Louis) where they contain copper, which has given rise to considerable works; 2dly, near Selbach, Gumbsweiler, Eckelhausen, and Ellweiler; 3dly, at Herrstein. These different positions are perhaps the traces of a porphyritic zone, which would be situated between the schistose rocks of the Hunsdrück and the trap of the Nahe: at Harrstein the porphyry is even inclosed between the schist beds; but its paste is here greenish, it contains crystals of hornblende, and perhaps belongs to a different formation. Porphyry constitutes much more extensive masses on the E. of the trap formation, which occur in the midst or on the edges of the coal basin of the Glane. The principal of these masses are, 1st, the Donnersberg or Mont Tonnerre; 2dly, the Königsberg, near Wolfstein, and 3dly, the group on the S. of Creutznach. 1st, The porphyritic formation constitutes the whole nucleus and a great part of the sides of the Mont Tonnerre, the mass of which is at least 8 or 10 leagues in circumference. This mountain is the most elevated point of the Palatinate: its height is about 600 metres [1968 feet] above the level of the Rhine at Mayence; it is as if isolated in the midst of a country of low hills, above which it rises more than 400 metres [1312 feet]. These hills are, on the N. and W. of Mont Tonnerre, formed of the coal measures; and to the E. and S. by micaceous red and slaty sandstone. Two leagues to the S. of Mont Tonnerre, a red sandstone of a different nature is found, mixed with quartzose conglomerate, which constitutes the last mountains of the Vosges. On the N. W. and S.W. branches of the trap formation are seen, which appear to rest against the foot of

Mont Tonnerre, in the midst of the coal measures, which very certainly cover the porphyry. There are proofs of the latter superposition in many pits sunk in the coal measures, between Kircheim-Boland and Orbis, on small veins of mercury, and in which porphyry has been met with at a slight depth. The porphyry of Mont Tonnerre is generally of a very pale and often almost white colour. The crystals of felspar, quartz, or mica, are not numerous, and the rock is often a nearly pure petrosilex; often also it apparently becomes almost entirely siliceous; sometimes on the contrary the texture of the paste becomes less close, and the rock passes into argilolite. I have not observed any true stratification in it. At Langenthal to the N.W. of Mont Tonnerre, a very regular and vertical vein of oxide of tin ore is worked in the porphyry. At Imschbach not far from thence, veins of silver, copper, and cobalt ores were formerly worked.

2dly. The Königsberg is less considerable; it is yet much more elevated than all the mountains of the coal measures with which it is surrounded. The rock constituting it resembles that of Mont Tonnerre, and like it is often a nearly pure petrosilex. It contains numerous veins of mercury and pyrites, accompanied with crystals of sulphate of barytes, which have been and still are worked at the mines named Wolfstein. On the sides of an adit level of these mines,* the petrosiliceous rock occurs as an assemblage of inclined prisms laid upon each other. Round Königsberg, highly inclined beds of blackish limestone are in some places seen, which rest on the porphyry, and dip beneath the coal measures that surround it. To the W. of Wolfstein, near Horschbach, the porphyry forms another elevated summit in the midst of the coal country of the Glane. It is again found on the banks of this river, near Ulmet, forming steep rocks, and the coal measures are seen to rest on it.†

* I have observed this prismatic disposition in the gallery of the mines of St. Elias, and Christians-Glück.

† I have not myself observed this fact; but I find it mentioned in the Geognostische Studien of M. Rheinheimer, p. 82.

3dly. At the salines of Creutznach, half a league on the S.E. from the town of the same name, near the village of Münster, the two banks of the Nahe are formed of steep rocks, 200 metres [656 ft.] high, of a porphyry analogous to that of Mont Tonnerre and Königsberg, but often affording varieties of a deeper colour, and in the paste of which a greater number of crystals are observed than in the other localities. Yet these crystals are occasionally of the same colour as the paste, and they are with difficulty distinguished. Sometimes also this paste appears to envelope elongated nodules, which are either of the same nature with it, or contain geodes of calcareous crystals. The porphyry there occurs over a great extent of country: on the S. W. it constitutes the mountains to within a league of the north of Obermoschel, with the exception of some interruptions in which the trap or coal measures occur; on the E. it occurs in the same manner to the environs of Wöllstein, and beyond Fürfeld: the paste is here a true argilolite; to the W. proceeding up the Nahe, the porphyry is followed for more than a league without interruption, and afterwards at different intervals; at Niederhausen, it presents traces of nearly vertical beds, which appear to dip beneath the trap rocks; near Bingert it constitutes the elevated mountain of Lemberg and is covered by the coal formation, in which, as at Orbis, small veins of mercury are cut at a slight depth by the petrosiliceous rock. On the N. at a short distance fram the *salines*, the porphyritic rocks cease on the right bank; a little further, they also disappear on the left bank, and the Nahe enters into a country of plains and low hills, formed of red sandstone, the beds of which evidently rest on the porphyry.

Notwithstanding the numerous escarpments that this great porphyritic mass presents, I could not observe any true stratification in it. The rock separates into decided prisms near Fürfeld. Copper mines have been worked in the porphyry near the salines of Greutznach. If we may judge from the number and extent of the subterraneous works occurring in this locality, the extraction ought to have been both considerable and productive. The principal vein contains pyritous and grey copper, which has been remarked to me as a circumstance occurring solely in the copper of the Palatinate situated in the porphyry, (Dippenweiler and Creutznach), whilst it is stated that the copper veins worked in the trap principally produce carbonate of copper.

Lastly, it is from the porphyritic rocks, on the banks and even the bed of the Nahe, that the numerous salt springs of the Creutznach rise, and borings through the bottom of the pits in which these springs are collected, have been driven more than 60 metres [about 197 feet] from the surface without meeting with any other rock than porphyry. This appears to me a very remarkable fact, and I consider it unique in the history of salt springs, which every where else rise from rocks of sediment. The springs of Creutznach are of a temperature a little above that of the mean temperature of the atmosphere; they contain only about a hundreth part of marine salt, and with this salt muriates of lime and magnesia and a little bitumen, but not an atom of the earthy and alkaline sulphates, which occur in all the saline springs of the E. of France and the N. of Germany.

To complete the notice of all the localities in which I have observed porphyry, I should add that the summit of the Landsberg, near Obermoschel, presents steep rocks apparently of the same nature, and which have also been thought to have been met with in the mines of that mountain, and in those of Stahlberg; but, as I have already had occasion to state, these two mountains are so much disturbed, and they contain, especially the former, so many different rocks, with so much disorder manifested in their union, that it appears to me no conclusion can be deduced from the observations to which they give rise, for determining the relative age of the different rocks of which they are formed. Abstracting therefore these two last positions, I shall cast a coup d'œil over the porphyritic rocks I have just mentioned.

Many geologists have apparently regarded them as constituting part of the trappean formation of the Nahe,* and

* M. Omalius d'Halloy, Journal des Mines, No. 144; M. Calmelet, Journal des Mines, No. 146, p. 142.

it cannot be dissembled, that in certain localities there appears to be a great connexion between the porphyry and trap. Small beds of argilolite have been cited at Schaumberg, (i. e. of the rock which, in some places, forms the paste of the porphyry), as appearing to alternate with the trap rocks.† I have moreover observed, in some trap rócks in the environs of Mont Tonnerre, rocks of a reddish colour, which might be offered as indicating a passage into porphyritic rocks; lastly the copper ore that has been worked in this porphyry, may also be noticed as indicating another resemblance between this rock and the trap, in which copper so frequently occurs, that its name has been given to it.

But on the other hand it should be considered that the porphyry and petrosilex have no where been observed in beds in the coal measures, or trap rocks; that they, on the contrary, form entire mountains or considerable masses of isolated rocks, and that it has been observed at Creutznach; that, independently of the great height of the escarpments which confine the valley, the porphyritic rock was, at the bottom of this valley, more than 60 metres [about 197 feet] thick, without the mixture of any other rock; that the position of these porphyritic masses with regard to each other, does not permit us to regard them as parts of thick beds interstratified with the other rocks; lastly, that in every place where a distinct superposition can be observed, the porphyry occurs beneath the red sandstone, the coal measures, and even the trap; and as it appears to me that more weight should be given to facts than to vague or probable inductions, I also consider that the porphyritic rocks should be regarded as very probably anterior to the coal measures and trap rocks, and consequently as constituting the most ancient of all the formations we are acquainted with in the Palatinate. If this be admitted, it becomes also probable that the different porphyritic masses situated on the E. of the trap band, belong to a single formation, on which rests the whole coal basin of the Glane.

The conclusions to which the preceding observations have

† Steinheimer, Geognostiche Studien, &c. p. 83.

led us, only making the geological singularity presented by the saline springs of Creutznach the more striking, I consider it interesting to add a few words on other saline springs situated in the eastern part of the Palatinate, but which offer with the springs of Creutznach, equally remarkable differences and analogies; I speak of the springs of Dürckheim, which also feed a very considerable *saline*, placed near the town of that name, at the foot of the eastern slope of the mountains of the red sandstone of the Vosges, and on the edge of the great valley of the Rhine, the surface of which is composed of more recent rocks. The position of these springs appears then, at first sight, to place them among common salt springs, and particularly those of Relchinge, of which mention was made at the commencement of this notice. The pits, in which the salt water of Dürckheim is collected, are cut across the sandstone, and yet this water bears a great resemblance to that of Creutznach: it is still more slightly salt than the latter, and contains less than a hundredth part of muriate of soda: it contains besides only earthy muriates and a little bitumen, without any trace of sulphates. These resemblances lead us to presume that the springs of Dürckheim rise from beneath the red sandstone, and probably from the same porphyritic rock as those of Creutznach.

I have above noticed the saline springs which rise from the coal measures, an uncommon position for these springs, though much less singular than that of the porphyritic rocks. I shall lastly notice, but solely for the sake of comparison with the known facts of Creutznach and Dürckheim, another spring of the same kind, also very weak, which flows over the ancient slates of the Hündsruck, near Brodenbach, on the banks of the Moselle, 4 leagues to the S. of Coblentz. M. Calmelet has noticed two others at Salzig near Boppart, and at Hoffelt near Barweiler. See Journal des Mines, No. 146. But not having visited these two localities, I know not from what rock these springs appear to rise.

On the Zoological characters of formations, with the application of these characters to the determination of some rocks of the Chalk formation. By Alexandre Brongniart, *Member of the Royal Academy of Sciences, &c.*

(Annales des Mines, 1821.)

Read at the Academy of Sciences, Sept. 3, 1821.

IN a report that I made to the Academy of Sciences in 1819, I was led to present a collection of facts I had brought together, in order to draw the attention of naturalists to the remarkable assemblage of circumstances which accompany each kind of rock in very distant countries, and under very different latitudes and meridians. These interesting resemblances, which had not as yet been offered, at least in so complete and evident a manner, which were in a great measure due to observations as yet unpublished, formed but a sketch then too little finished, for publication.

Bnt I have since then taken up some of the subjects contained in this general sketch, and I have endeavoured to develope them, and offer proofs sufficient to confirm the results.

The organic remain that was the cause of these remarks, was a trilobite sent from North America by M. Hosach: this remain of an animal of the crustaceous class, presented a species and position resembling those observed in Europe.

To the use that may be made of organic remains in the determination of formations I now return, by applying it to another class of rock, to one which forms part of our country, but which, more ancient than that which forms the surface, often dips beneath it. This rock is the chalk. My object is to shew the organic remains contained in it offer characters sufficient for recognising it in situations very distant from each other, when those drawn from its consistence, stratification, colour, &c. have disappeared, and when its superposition is either obscure, uncertain, or difficult to be recognised.

It should be recollected, that the mass of rock referred to the chalk formation, is divided into three sub-formations; the upper or white chalk; the middle, which is grey chalk or craie-tufau; and the inferior, or chalk mixed with green grains, which M. Berthier has determined to be silicate of iron (fer silicaté) with water, and which I shall name, considering it as a mixed rock, *glauconie crayeuse*. It is the green sand of the English geologists.)*

These three divisions of chalk contain fossil organic remains, which are generally different in each sub-formation, but there are at the same time some which are common to all.

§ I. *Value of Zoological characters in geology.*

Among the different chalk rocks I am about either to cite or describe, many will without difficulty be regarded as belonging to this formation; some are even generally recognised to form part of it; to these last I shall only add zoological proofs to the geological resemblances that have already been established.

I shall also refer this formation to places in which until lately, chalk has not been recognised, in which this rock is so much disguised, that I shall have some difficulty in causing its analogous formation to be admitted with the inferior or chloritous chalk (green sand), to which I consider it may be associated. In one of these situations the mineralogical characters entirely disappear, the geological position is obscure, and the zoological characters alone remain.

* From this it would appear that the upper or white chalk of M. Brongniart, is our flinty or upper chalk, with most probably our *white* chalk without flints or lower chalk; that his midd' chalk is our grey chalk, and perhaps chalk marl, and that his inferior chalk or glaucoine cray-euse, is our green sand. (Trans.)

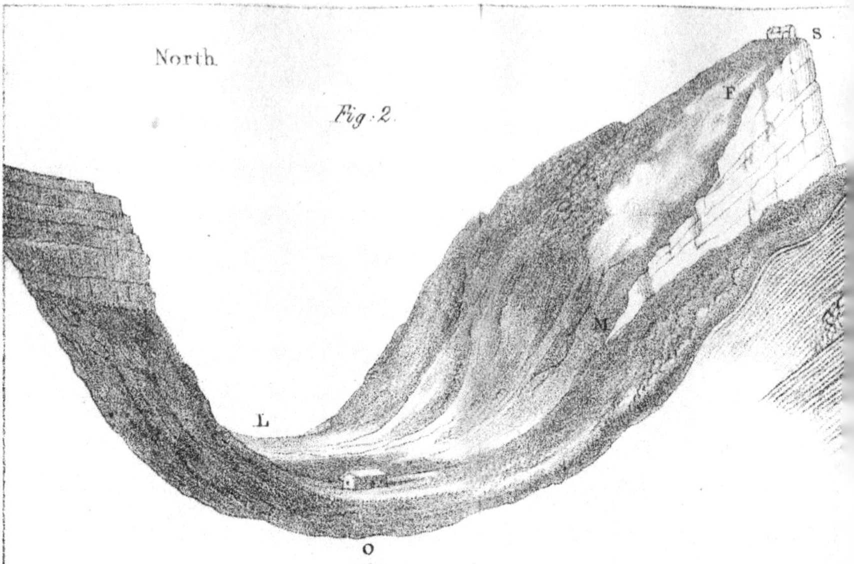

North
Fig: 2.
S
F
M
L
O

West
Fig: 1
AB
m
M

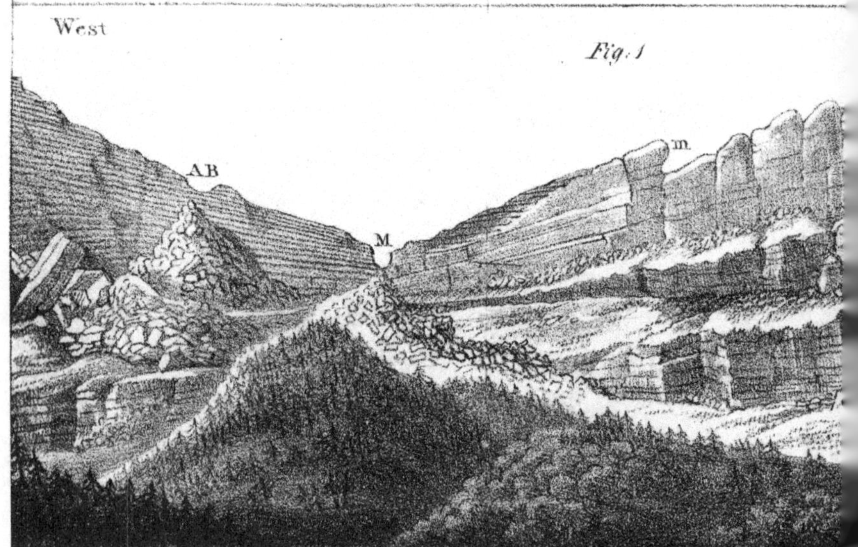

G. Scharf Lithog:

View

MONTAGNE DES F

London Pu

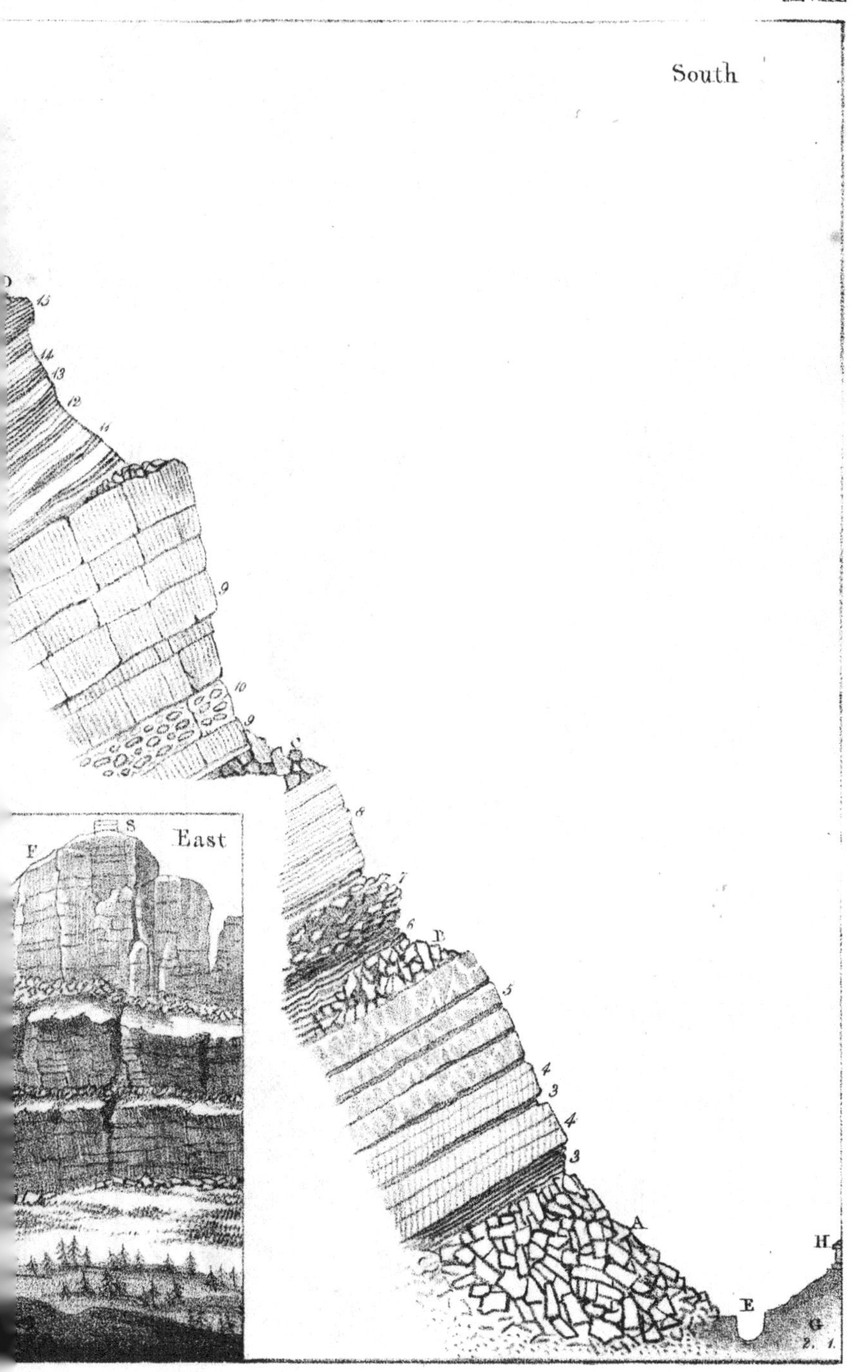

LLEY OF SERVOZ. Section

Printed by C. Hullmandel.

illips. 1823.

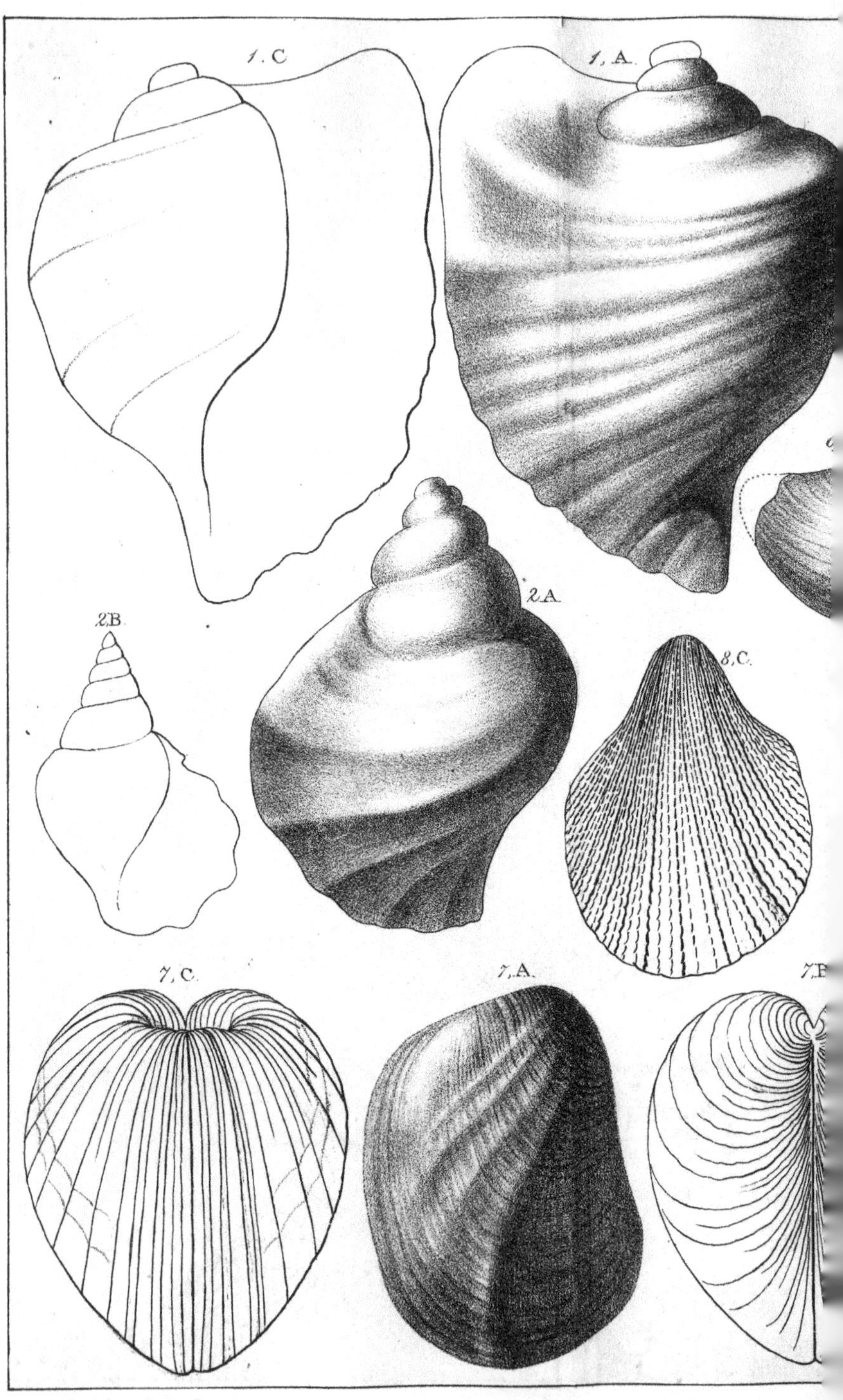

G. Scharf Lithog:

London. Publishe

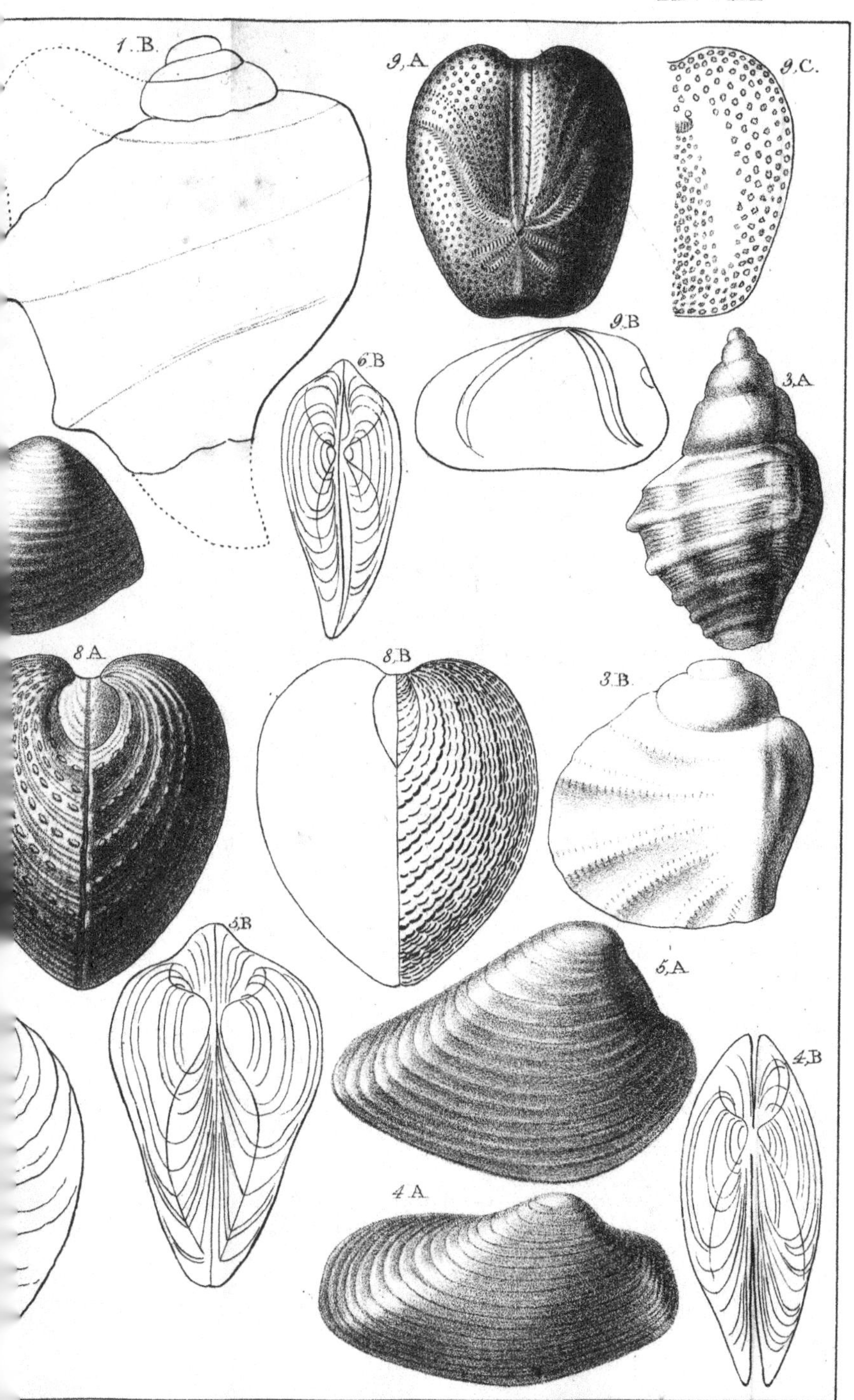

Printed by C. Hullmandel.

W. Phillips. 1824.

Before then I employ them alone and as most important, their value should be again examined.

It had long been remarked that differences were almost always found between the shells that now live in the seas, and those found fossil in all countries. This has been confirmed by a more detailed examination, and has gradually led to another rule, that the deposites of organic remains buried in the beds of the globe have been formed, as it were, in successive generations, so that all the debris of the same deposite possess with each other a certain sum of resemblances, and with superior and inferior deposites a general sum of differences; it has been also thought that the latter sum becomes larger, or the differences greater, the more these deposites are distinct, or at greater distances from each other in a vertical direction. This rule, at first laid down timidly, and only for certain localities (as should always be done when rules are to be established which can only result from the observation of numerous facts)—this rule, I say, may apparently be applied to all places observed in different parts of the globe, and to all the organic remains buried in its beds, whether they belong to the animal or vegetable classes. Up to the present time, the objections that have been offered have either vanished before a more scrupulous examination, or have been explained by the discovery of peculiar circumstances that have given rise to them. This rule, reduced to the general exposition of it we have just given, does not appear to be susceptible of any real objection, and all geologists are now agreed that the generations of organized bodies which have successively inhabited the surface of the earth, differed the more from the present, in proportion as their remains were found buried deeper in the beds in the earth, or what is nearly the same, as they lived at epochs farther removed from the present.

Consequently if no other difference were observable in the beds which compose the crust of the globe than the distinct succession of generations of organic bodies, that alone would be sufficient to establish (as has been remarked by M. Cuvier) that this crust has been formed in successive depositions, at different periods.

But this character of succession in the formation of the beds of the earth, is frequently associated with other remarkable differences, such as the nature of the rocks, their structure on the great scale, their acknowledged order of superposition, the minerals that accompany them, &c.: now these mineralogical differences are almost always in accordance with the characters derived from the general resemblance of the organic remains in those deposities, which are regarded as of the same formation from their geological characters, and they are found very generally in accordance with their differences in the inverse case.

Yet there are cases in which these two classes of characters, without being in manifest opposition, no longer accompany each other.

These cases occur in the two formations I am about to refer to the chalk. The question then is, to which of the two characters we should give the preference, in order to determine the epoch of formation of the rock, which no longer presents them associated together, that is to say, to answer the following question:

"When in two rocks, far distant from each other, the rocks themselves are of a different nature, whilst the organic remains are analogous, should we, from this difference, regard these rocks as of a different formation, or should we, from the general and *properly determined* resemblance of the organic remains, consider them of the same epoch of formation, when the order of superposition does not evidently oppose itself to this conclusion?"

We must not forget that one of the principal objects of geology, is to distinguish the different epochs which have succeeded each other in the formation of the globe, and to determine which are the rocks that have been formed at nearly the same epoch. Now, it will be acknowledged that rocks of very different natures may be formed at the same time, almost at the same moment, not only in different parts of the globe, but also in the same place.

We cannot but admit a conclusion drawn from facts which we have under our eyes; for all that now occurs on the surface of the globe certainly belongs to the same geological epoch, which commenced at the moment our continents took their present forms; and although this epoch does not furnish instances of geological phenomena on a grand scale, and we meet with but few cases of the formation of new rocks, yet they are sufficient to shew us, for example, that the argillo-trap rocks formed by Vesuvius, and most of our volcanoes, the calcareous rocks formed by many of our springs, and the siliceous rocks formed by some others, (those of Iceland, &c.), are mineralogically very different from each other; but that the organic remains they envelope have all the common character of the generation established on the earth since the commencement of this epoch.

It is not the same with respect to the generations of organized beings: they may be destroyed in an instant; but it certainly requires considerable time to recreate them so that they be developed in the number and variety in which they are usually presented to us. This developement supposes a long series of ages or at least years, which establish a true geological epoch, during which all the organized bodies which inhabit, if not the whole surface of the globe, at least very extensive parts of that surface, have acquired a certain character of family or epoch, which cannot be defined, but which cannot be misunderstood.

I then consider the characters of *the epoch of formation* drawn from the analogy of organized bodies as of the first importance in geology, and as superior to all other differences, however great they may appear.

Thus when the characters derived from the nature of the rocks, and what is of least value, of the height of the rocks, of the formation of the valleys, even the dip of the beds and the greatest non-comformable stratification, are opposed to that which we derive from organic remains, I should still give the preponderance to the latter;—for all these circumstances, all these differences may result from an instantaneous revolution or formation, which does not, in geology, establish a special epoch. Without endeavouring to prove this principle by longer discussion, it will be sufficient to cite one fact: The rocks of Calabria have been, during thirty-eight years, the theatre of terrible disturbances; horizontal beds have been set on their edges, entire masses of rocks have been transported to a distance, and have been placed in unconformable stratifications on other rocks, and no geologist has proposed to consider these masses and rocks as of a different geological epoch. Circumstances of greater value, much more general phenomena, and periods of longer duration, are required for the change of organized beings. The rocks of Calabria suffered, in a short time, derangements which may be compared to those observed in the beds of the Alps, but during five or six thousand years organized species have not manifested any appreciable change in their form and other qualities.

Yet I do not pretend to state that characters derived from the relative disposition of beds (but not the evident superposition) (superposition evidente), their nature, &c. should not be employed, even with confidence, by geologists, to determine different epochs of formation; alone or united with those derived from the organic remains, they are of the greatest value; but I consider, and I think that I have good reasons for this opinion, that when these characters are opposed to those drawn from the presence of organic remains, the preference should be given to the latter.

It must be confessed that much attention and caution should be used in the employment of this character. I am not ignorant that the influence of horizontal distances or climates on specific differences should be distinguished and estimated; that we should learn to appreciate the apparent and sometimes real resemblances that some species present in very different formations, and which have possessed the rare privilege of surviving the destruction of their contemporaries; and to continue the same amidst all the changes which have taken place around them:—I am not ignorant that we should also learn to recognise the individuals torn from other rocks, and transported by certain causes into newer rocks, and to distinguish them from those which have lived on the spot, and at the period when the species to which they belong were characteristic.

I am aware of all these difficulties: I am on my guard against these causes of deception, which introduce the uncertainty and difficulty which we meet with in other sciences, into geology, and which require continued attention and labour from the geologist, so as with discernment to choose the species from which he should derive his characters, and attach their true value to them.

I have then examined with all the attention that circumstances have permitted, the influence of these different causes on the structure of the chalk rocks, of which I am about to speak.

These rocks are so singular that I considered it necessary to precede their description by the general considerations above brought forward, and to prepare naturalists for recognising as chalk a hard and black rock which occurs at a height of more than 2000 metres [6560 feet] on the summit of a mountain so difficult of access at certain periods, that I have been unable to attain the point where it is found.

But before I proceed to the determination of this singular chalk, I shall examine some others, whose differences being less singular, will lead us less suddenly to that with which I shall terminate this notice.

§ II. *Chalk of Rouen, of Havre, and of the Coast from Honfleur to Dives.*

A hill named St. Catherine is observed near Rouen, at the eastern entrance to the town; this steep hill presents the union of the upper white chalk with the craie tufau and the inferior chlorite chalk (green sand), and leaves no doubt as to the identity of formation of these two rocks; but these last contain a great quantity of organic remains, which differ from those found in the white chalk. This union of circumstances is very favourable for observation, and we present it the first, because it affords us the means of referring rocks to the chalk formation, which at first sight appear very different from it. Thus only the two last chalks are seen at Cap de la Heve, near Havre, at Honfleur, &c.* This inferior chalk is the same with that observed in England by De Luc, between Beachy Head and Sea Houses, on the coast of Sussex, and so well described by that geologist.† This chalk does not differ from that which Mr. William Phillips has observed to the W. of Calais on the coast of France, between Sangatte and St. Pot; and which appears to correspond exactly with that of the English coast between Dover and Folkstone. In both these situations, as in many other places, the white chalk and craie tufau are separated from the glauconie craieuse (green sand of the English geologists) by a bed of bluish clay marl.‡ Among the fossil shells which occur in these chalks and which apparently characterize them, I shall mention the following, as coming from the three places I have just noticed, i. e. Rouen, Havre, Honfleur, and even the continuation of this coast to Dives.

Organic remains of the craie tufau and green sand (glauconie craieuse) of Rouen, Havre, Honfleur, the environs of Dives, &c.

Nautilus simplex ... Sow:	....	Rouen.
Scaphites obliquus.. Sow:	(pl. viii. fig. 4.) ‖ ...	Rouen, Brighton.
Ammonites varians . Sow:	....	Rouen, Havre, where it occurs of a large size.
—— inflatus ...... Sow:	....	Havre.
—— rothomagensis. Defr:	(pl. vi. fig. 2.)	Rouen, it acquires the size of more than a decimetre.

* I have myself observed the structure of the hill of St. Catherine, and of the cliffs from Honfleur to Dives; but I am indebted to M. St. Brice for a great part of the shells from the former place. All I know concerning the structure of Cap la Heve I owe to M. Audouin.

† Geological letters to Blumenbach, p. 200.

‡ Transactions of the Geological Society.

‖ The Plates quoted in parentheses are those of the new edition of the Description geologique des environs de Paris, in which part of this memoir is inserted.

The initials Defr. represent Defrance.
A. Br. .... A Brongniart
Sow. .... Sowerby.
Lam. .... Lamark.

Species	Authority	Plate	Locality
Ammonites Coupei	A. Br:	(pl. vi. fig. 3.)	Rouen.
—— Gentoni	Defr:	(pl. vi. fig. 6.)	Rouen.
Hamites rotundus	Sow:	(pl. vii. fig. 5.)	Rouen.
Turrilites costatus		(pl. vii. fig. 4.)	Rouen, Havre.
Turbo?	Undecided interior casts		Rouen.
Trochus	Interior casts which may apparently be referred to the trochus of the Perte du Rhône known by the names of T. Gurgitis, Rhodani, Cirroides. (pl. ix. figs. 7. 8. 9.)		Rouen.
Cassis avellana	A. Br:	(pl. vi. fig. 10.)	Rouen.
Podopsis truncata	Lam:	(pl. v. fig. 2.)	Rouen, environs of Tours?
—— striata	Lam:	(pl. v. fig. 3.)	Havre, Brighton.
Inoceramus concentricus		(pl. v. fig. 11.)	Rouen.
Ostrea carinata	Lam:	(pl. iii. fig. 11.)	Havre.
—— pectinata	Lam:		Havre.
Gryphæa columba	Lam:	(pl. vi. fig. 8.)	Havre, Blanc Longleat.
Pecten quinquecostatus.	Sow:	(pl. iv. fig. 1.)	Havre. It appears slightly to differ from that of the white chalk
—— intextus	A. Br:	(pl. v. fig. 10.)	Havre.
—— asper	Lam:	(pl. v. fig. 1.)	Havre.
—— dubius	Defr:	(pl. iii. fig. 9.)	Rouen.
Plagiostoma Mantelli	A. Br:	(pl. iv. fig. 3.)	The coast of Dover.
Plagiostoma spinosa	Sow:	(pl. iv. fig. 2.)	Rouen. It occurs in the craie tufau close to the white chalk, and does not appear to differ from the species which belongs to the latter rock. Brighton, &c.
Trigonia	An interior cast which apparently indicates a species near Tr. scabra. Lam: or Tr. striata. Sow: which may be the same species, (pl. ix. fig. 5.)		Rouen.
Mytiloïdes? labiatus	Ostracites labiatus. Schlotheim, (pl. iii. fig. 4.)		Rouen, & in the craie tufau of many other places.
Crassatella	Interior casts which apparently indicate small species of this genus.		Rouen.
Terebratula semi-globosa	Sow. Lam.	(pl. ix. fig. 1.)	Rouen, Havre.
—— gallina	A. Br:	(pl. ix. fig. 2.)	Havre.
—— alata	Lam:	(pl. iv. fig. 6.)	Havre.
—— pectita	Sow:	(pl. xi. fig. 3.)	Havre.
—— octoplicata		(pl. iv. fig. 8.)	Havre.
Cidarites variolaris	A. Br:	(pl. v. fig. 9.)	Havre.
Spatangus Bufo	A. Br:	(pl. v. fig. 4.)	Havre.
—— suborbicularis	Defr:	(pl. v. fig. 5.)	In the craie tufau of the environs of Dives.

Q 2

§ III. *Chalk of the environs of Perigueux and Bayonne.*

The chalk has been observed to end, on the south of Paris, at the southern confines of the department of the Indre. It really ceases there, for the rocks which succeed it are inferior to it; but when, proceeding to the S.W. these older rocks are passed over, the craie tufau is again found in the department of the Dordogne, in the environs of Perigueux, and especially on the west of that town. The high and steep hills which border the river Lillefrom Perigueux to the place named La Massoulie, are composed of sandy and often micaceous grey chalk, i. e. of craie tufau which occurs in an immense mass without any distinct strata throughout the greater part of its extent; stratification is however indicated by seams of black flints which divide it into numerous beds. These flints, which belong rather to the variety we have named silex cornés (splintery hornstone) than to true flints, are, as we have elsewhere stated, characteristic of the *craie tufau*, into which they seem to pass.

The shells which this chalk contains are numerous in some places, and although I observed this hill very rapidly,* I was able to collect the following species:

List of some shells of the craie tufau in the environs of Perigueux.

Nautilus Pseudopompilius.	Schlotheim.	
Trochus		Indeterminate interior casts.
Ostrea vesicularis.........	Lam. (pl. iii. fig. 5.)	The individuals are smaller, & resemble those of Luzarche, & the latter the small individuals of Meudon.
Gryphæa auricularis......	A. Br. (pl. vi. fig. 9.)	
Plagiostoma spinosa	Sow. (pl. iv. fig. 2.)	Although I have only seen the interior surface of a few valves I do not doubt the correctness of this determination.

* I visited this place in 1808, and then observed the characters of the craie tufau formation.

Proceeding still further south, the chalk formation is found in places where it has not until now been noticed. I do not hesitate to refer the grey, hard, sandy, and micaceous rocks which form the base of ground in the environs of Bayonne, and especially the coast and rocks of Biaritz, to this formation. It was in 1808 that I conceived this idea on the epoch of formation of this rock. The subsequent examination that I made of its accompanying circumstances, of its resemblance to some varieties of the craie tufau, and the spatangus that comes from the environs of Bayonne, which I have named, from M. Defrance, spatangus ornatus, and of which I have given a figure pl. v. fig. 6. of the Description geologique des environs de Paris, fully confirm me in the opinion that this rock should be referred to the craie tufau. It occurs as a continuous mass, in which distinct stratification cannot be recognized but by means of the different degrees of solidity of the parts composing it; in fact alternating zones are observed in it of a greyish, argillaceous, or sandy limestone, easily decomposed, and a hard limestone, as it were divided into a suite of irregular nodules, which project on the surface of the escarpments in the same manner as the flints in the white chalk.

This mass contains numerous remains of fossil shells, which it was impossible for me to determine, but among which I recognized echinites, the spatangus ornatus, and sp. bufo, to occur. I did not observe any ammonite.

Notwithstanding the specific differences which many of these shells bear to those of the chalk, the mass of resemblances teaches us that they approach the species that then existed, more than those of any other epoch.

§ IV. *Chalk of Poland.*

I shall, out of the vast extent of chalk in Poland, select three points at a distance from each other, on which I possess certain data.

The two first are in the environs of Grodno in Lithuania,

and of Krzeminiec in Volhinia. The chalk is there white like that of Meudon; it contains like it black flints, belemnites, (but apparently of a different species), cidarites vulgaris, plagiostoma spinosa of Sowerby, and probably other organic remains, which the few specimens we possess do not permit me to recognise.

It appears that all the chalk of Poland presents the same resemblance; for Mr. Buckland, who observed it in place, wrote me in 1820, that, "the chalk on which the castle of Cracow stands, is precisely similar to that of Meudon, full of echini and flints: it is perhaps a little harder. I have not seen the plastic clay in contact, but I have observed shells in the collections of Cracow, which resemble those of the calcaire grossier, and sub-appennine mountains, which were said to have been found at a short distance to the N.E. of Cracow; I have no doubt on the identity of the two formations."

§ V. *Green Sand of the Perte du Rhône near Bellegarde.*

Two very different rocks are observed at this remarkable place: the inferior is a compact, fine, grey, and yellowish limestone, disposed in regular and nearly horizontal strata, which at first sight shew no organic remains. De Saussure observed this circumstance, and assures us that none were ever found in it; extensive and numerous cavities occur in this compact limestone, through which the waters of the Rhone are precipitated.

Between these beds, and probably even beneath them, marl beds occur, as throughout the Jura, very different from the limestone I have just mentioned, and which contain a great quantity of shells. I am only acquainted with these fossil shells from the statement of M. Deluc, and the specimens he has sent me; but the species of these fossils, and the nature of the accompanying stone, establish the greatest resemblance between these marl beds and those interposed in the midst of the Jura limestone.

This rock appearing from its position altogether foreign from that which principally occupies me, it will suffice to notice these shells by a name and a figure, so that we may possess the means of describing these characteristic shells; but it is my intention to unite the history of their association, and their description, with that of the shells which belong to the Jura limestone, and which will be the subject of another work.

The organic remains which I shall content myself by noticing in this memoir as characterizing this formation in the Jura, and many other places, are the following:

Strombus Pelagi ...	A. Br. (pl. 8 of this work* fig. 1. A. B. C.)..	The Perte du Rhône.
——— Oceani	A. Br. (ib. fig. 2. A. B.).	In an argillaceous marl absolutely the same as to colour & position, at Cap la Hêve near Havre, & in the Jura.
—— Ponti	A. Br. (ib. fig. 3. A. B.).	With the preceding.
Cardium Protei	A. Br. (ib. fig. 7. A. B.).	In the same places.
Hemicardium tuberculatum.........	A. Br. (ib. fig. 8. A. B.).	In an argillaceous marl at Cap de St. Hospice near Nice, & at the Perte du Rhône.
Mya? or Lutraria? Jurassi. ....:	A. Br. (ib. fig. 4. A. B.).	In the marls of the Perte du Rhône, and in the upper oolitic or compact limestone of the Jura; of Ligny, dep[t]. of the Meuse; of Soulaine, dep[t]. of the Aube; of Gondreville, near Nancy, &c.
Donacites Saussuri ...	A. Br. (ib. fig. 5. A. B.).	The Perte du Rhône.
—— Alduini	A. Br. (ib. fig. 6. A. B.).	Cap la Hêve.
Spatangus oblongus...	Deluc. (ib. fig. 9. A.B.C.)	From the Perte du Rhone & the argillaceous marls of the same position which are behind the town of Neufchatel.

The greater part of these shells are interior casts, but which have very well preserved their forms and characteristic salient parts, so that they may be determined with sufficient exactitude. There are also found in the same rocks belemnites, ammonites, trochi, terebræ, serpulæ, smooth and striated terebratulæ, &c. the complete enumeration and exact determination of which would carry me too far from the principal object of this memoir.

The greater part of the preceding shells are from the place named the Perte du Rhône; but they are not from the rock analogous to the inferior chalk (green sand) which rests upon it.

This second rock, which is above that we have just noticed and to a certain degree characterised, possesses a very distinct and nearly horizontal stratification, dipping slightly to the S.E.; the thickest lower bed is composed of a yellowish limestone, often shaded or veined by yellowish argillo-ferruginous portions: it appears composed of an immense mass of lenticular stones, which were at first taken for nummulites), or a multilocular shell, but which have since been recognised to be small madrepores, to which M. de Lamarck has given the name orbitolites lenticulata. Above are alternating strata of marly limestone and sandy clay mixed with the green grains which are constantly found in the lower parts of the chalk.

This rock contains numerous organic remains, the resemblance of which to those of the green sand (craie chloritée) struck me the first instant I saw them. This resemblance long since struck M. Deluc (the nephew), and he remarked it to me when we examined in his collection the numerous organic remains of this rock, which have been assembled by his uncle and his father. The analogy is still more complete and apparent, when, as he has done, these fossils are placed by the side of those of Folkstone in England, which come from the green sand; but they were still more decisive when I was enabled to compare these shells with those from Mount St. Catherine near Rouen. Yet these relations are more real and easier to seize from their general features than from a particular examination of these bodies. Thus nearly the same genera are found in these three situations, and the species so resemble each other that to perceive their difference they must be placed close together; some species are perfectly identical. The following comparative list, derived from the shells which I obtained in great numbers on the spot, and from those collected by Messrs Deluc since 1750, which have been obligingly sent me by M. J. A. Deluc, will suffice to give a precise idea of these relations.

Organic remains of the green sand of the Perte du Rhône, near Bellegarde:

Belemnites	Undeterminable.		
Ammonites inflatus	Sow.	(pl. vi. fig. 1.)	From Rouen & Havre, it greatly varies in size. The increase of the last whorl is not very sensible in small individuals.
—— Deluci	A. Br.	(pl. vi. fig. 4.)	
—— canteriatus	Defr.	(pl. vi. fig. 7.)	Collection of Deluc.
—— subcristatus	Deluc	(pl. vii. fig. 10.)	Collec. of Deluc. It very much resembles A. cristatus from Folkstone.
—— Beudanti	A. Br.	(pl. vii. fig. 10.)	Coll. of Deluc.
Hamites rotundus	Sow.	(pl. vii. fig. 5.)	From Rouen. These are orthoceratites of De Saussure.
—— funatus	A. Br.	(pl. vii. fig. 7.)	Coll. of Deluc.
—— canteriatus	A. Br.	(pl. vii. fig. 8.)	Coll. of Deluc.
Turrilites Bergeri	A. Br.	(pl. vii. fig. 3.)	Coll. of Deluc.
Trochus Gurgitis	A. Br.	(pl. ix. fig. 7.)	Coll. of Deluc.
Trochus? Rhodani	A. Br.	(pl. ix. fig. 8.)	It is also found at Lignerolle above Orbe.
Trochus? Cirroïdes	A. Br.	(pl. ix. fig. 9.)	The cast of this shell occurs in the chalk at Rouen, Havre, and Brighton.
Cassis avellana	A. Br.	(pl. vi. fig. 10)	At Rouen also. They are at first sight taken for turbines and ampullariæ.
Ampullaria?	An indeterminable interior cast.		
Eburna!			
Cerithium excavatum	A. Br.	(pl. ix. fig. 10.)	Coll. of Deluc.

Gryphæa aquila	A. Br. (pl. ix. fig. 11.)	Coll. of Deluc. These are the shells noticed as oysters by De Saussure. I consider this gryphite as of the same species as that found near Rochelle, (fig. 11. C,) in a rock which also bears a great analogy to that of the craie tufau.
Pecten quinquecostatus	Sow. (pl. iv. fig. 1.)	Coll. of Deluc. At Rouen, Havre, & all chalk rocks.
Lima or Plagiostoma pectinoides	Sow.	Coll. of Deluc.
Spondylus? Strigilis	A. Br. (pl. ix. fig. 6.)	Coll. of Deluc.
Trigonia rugosa?	Lam. Park. Organic rem. vol. iii. tab. 12, fig. 11.	Coll. of Deluc.
—— scabra	Lam. Enc. pl. 237. fig. 1. (pl. ix. fig. 5.)	Coll. of Deluc. Rouen.
Inoceramus concentricus	Park. (pl. vi. fig. 11.)	Folkstone and Rouen.
—— sulcatus	Park. (pl. vi. fig. 12.)	Folkstone.
Lutraria Gurgitis	A. Br. (pl. ix. fig. 25.)	This species is well characterized, & differs from that found in the marls of the Jura limestone.
Terebratula Gallina	A. Br. (pl. ix. fig. 2.)	Coll. of Deluc.
—— ornithocephala	Sow.	
Spatangus lævis	Deluc. (pl. ix. fig. 12.)	Coll. of Deluc.
Cidarites variolaris	A. Br. (pl. v. fig. 9.)	The same with those found in the green-sand of Havre.
Orbitolites lenticulata	Lam. (pl. ix. fig. 4.)	Lenticular stone of the Perte du Rhône.

This last gives us, as we have said, numerous shells of the chalk epoch, but it does not afford us any shell either of the inferior or older rocks, or of the upper and newer rocks.

These considerations are sufficient to lead us to the conclusion, that the green sand resting on the Jura limestone at the Perte du Rhône belongs to the inferior chalk formation; that this rock, analogous to the green sand of the English geologists, as they themselves admit, is seen almost immediately in contact with the fine compact Jura limestone, and that it is only separated from it by an argillaceous marl, a disposition analogous to that observed in France, at Cap la Hève, Honfleur, and Dives, &c. and in England at Tetsworth, &c. We might increase the number of analogies by comparing the lenticular rock, penetrated by oxide of iron, which de Saussure describes as a true iron ore, to the iron sand (sable ferrugineux), which often occurs in beds of greater or less thickness beneath the green sand.

Thus, notwithstanding the very considerable distance of the places; notwithstanding the different form of the mountains and rocks; notwithstanding a few mineralogical differences, the yellowish calcareo-ferruginous rock mixed with greenish grains of the Perte du Rhône, presents, it may be said, a complete analogy to the green sand of the north of France and the S.E. of England, for the characters of the associated rocks, minerals, and superposition, agree with those afforded by the organic remains in establishing this analogy of formation.

§ VI. *Formation of the chalk epoch in the Buet chain, and a comparison of this rock with that of the transition class in the Montagne des Fis.*

We are now arrived at an approximation that appears much more extraordinary—one that I should still offer with hesitation (for my mode of considering this subject dates as far back as my Swiss tour in 1817), if my opinion had not been confirmed by that of Mr. Buckland, an opinion which that geologist entertained when he passed through Paris in the end of 1820, and which he has published in the Annals of Philosophy, for June, 1821.

A chain of elevations extends from the summit of the Buet (in the Savoy Alps) which apparently constitute a dependence on it, and which are remarkable for their black colour, for their form, which is often precipitous on one side, with a more or less rapid slope on the other, for their very considerable elevation above the level of the sea, an elevation which amounts to 2500 metres [8200 feet].

The principal mountains to which the following statement may be applied, are the Montagne de Varens, the Dent de Morcle, the Montagne de Sales, and the Rocher des Fis in the valley of Servoz:* I shall speak principally of the latter.

The Montagne des Fis, crowned by the rocks of the same name, precipitous towards Servoz for a considerable part of its height, and covered by the debris of the upper parts, is composed of numerous beds, which, from Servoz, appear nearly horizontal, because they incline from S.E. to N.W. The rocks forming these beds are calcareous and schistose, mixed with hornstone and flinty slate; they belong, as I have elsewhere stated,† to the transition class. But the description of the rocks composing this mountain having been foreign to the memoir in which I mentioned them, I omitted it. I shall therefore insert it here, in order better to expose the mineralogical and zoological differences that are perceptible between the base or lower part of this mountain, which belongs to the transition series, and the upper portion, which I refer to the chalk formation.

The Montagne des Fis, from its base in the valley of la Diosa, opposite Servoz, to that part of its summit which I visited, is composed of micaceous clay slate, compact limestone, of a blackish or deep smoke grey colour, and of different varieties of sandstone (psammite); but these rocks are covered, on many points of the southern face of the mountain, by immense fallen masses, which increase daily; these masses of debris conceal part of the beds which compose the mountain, and the cause which produced them appears to have been sufficiently powerful to allow entire masses of rocks to slide down without perceptibly deranging their structure, so that it would require a scrupulous and long examination of this mountain, to be certain that the masses or escarpments observed in the ascent, really follow each

* I visited these rocks in company with M. Lainé, in 1817.

† Memoir on the relative position of the serpentines &c. in the Apennines.

other in the order of superposition, as they successively present themselves; and that they are not portions of the upper parts which have slid down, and have been placed before the lower portions, and finally to be sure that we do not give as two series of beds that which only forms one.

It was necessary to state the above in order that the following enumeration may not be regarded as a perfect list of the succession of the beds. But even with all the inaccuracy that may be attributed to it, it will not the less present the different rocks which compose the southern side of the Montagne des Fis, and the *comparative results* which we may derive from them.

The hill (G) opposite this mountain, that on which is situated the village du Mont (H), and which is separated from the body of the mountain by a small valley in which the Nant de Siouve (E section; pl. 9) flows; a hill which may either be the base of the mountain, or independent, and the result of the slide of one of its parts, is composed of black greywacké slate (No. 1), containing spherical nodules. These nodules, which are a little harder than the rest of the rock, sometimes contain impressions of ammonites in their interior. (No. 2). That which I found was too incomplete to determine the species to which it belonged, but it appeared to me very different from the ammonites of the summit of which I shall presently speak, whilst it bears a great resemblance to the ammonite I have mentioned (in the memoir above cited) which comes from an alpine mountain of Oberhasli, of an analogous nature to this transition portion of the Montagne des Fis.

The beds of this slate dip towards the N. as do those of the mass of the mountain. After having passed this species of out-work and the small valley which separates it from the Rochers des Fis, we arrive at the foot of the rocks, and the precipitous escarpments they present, which would render this side of the mountain inaccessible, but for the fallen masses which have there formed very steep slopes, composed of large pieces of rock, always ready to fall, but which can however lead us to the summit.

When we have passed the first and largest collection of fallen masses, which descended in 1751, we observe in the first escarpment we pass:

1st. A micaceous clay slate (No. 3), resembling that of the hill du Mont; it is blackish and solid.

2dly. A greyish compact limestone (No. 4), which alternates with the slate, forming more rounded and solid salient portions.

3dly. Towards the upper part, a grey and blackish compact limestone (No. 5), full of numerous veins of spathose limestone, and also alternating with the slate (No. 3).

A second collection of fallen rocks is then traversed, and we reach another escarpment, which is composed,

4thly, Of a thick bed of argillaceous slate (No. 6), very fissile and splintery, traversed throughout, and in every direction, by nodular seams of white spathose limestone (No.7) and veins of quartz.

5thly. Of a schistose sandstone of a very compact texture, and yet of a very fissile structure (No. 8).

A third collection of fallen rocks (C) then follows, in a great measure composed of large parallelopipedal pieces, which are very extensive and difficult to traverse. It leads to a third escarpment, presenting nearly vertical walls of great elevation, the strata of which are cut into very large parallelopipeds, by fissures perpendicular to each other, and lined with calcareous spar. It exposes at its base,

6thly. A very compact greyish limestone (No. 9), traversed by steatitic and chloritic veins, disposed in the form of net-work, and in some strata constituting an amygdaloïdal steaschist (steaschiste amygdalin), the almond shaped portions being very large. This calcareous mass is lined with calcareous spar of the variety with the short dodecahedral prism.

7thly. Above, and in thick beds that are apparently solid, but which are however very splintery, is a micaceous and quartzose sandstone, alternating without order with the micaceous argillaceous slate and with the blackish schistose sandstone, so as to form very singular striped rocks (Nos. 11, 12, 13, & 14.)

8thly. Lastly, these sandstones are covered by a micaceous argillaceous slate in thin laminæ and very splintery, (No. 15), and apparently yellowish, but this colour is only superficial. It is blackish in the interior of the smallest pieces, and so fissile, so easily broken, and with such a want of cohesion, that notwithstanding the great mass it presents, a good sized specimen cannot be obtained.

This slate forms the sharp and steep crest AB of that part of the Montagne des Fis which we reached; but it is not the most elevated portion, the beds dipping towards the N.W. and their southern section presenting an inclination towards the W: this slate does not rise to the summit of the Pointe des Fis. This point, as well as that named Le Marteau (m. fig. 1.) is, as we have been assured, composed of limestone mixed with sandstone, and appears to be the prolongation of the beds No. 9 to 14.

It is on the back of the Montagne des Fis, and nearly at the point marked F, that the bed, containing the fossil shells I shall enumerate, occurs on a rapid and elevated slope, almost always covered with snow. The snow which covered it prevented me from visiting it, but M. Beudant, who reached this part of the mountain by the valley of Sales (Lo), in 1818, saw a portion of it in situ. He observed the shelly black rock nearly in the position given in the section. It is, according to him, in nearly parallel stratification with the transition rocks on which it rests. This compact limestone, which is hard, coarse, blackish, leaving on the surface of a solution of it in nitric acid, a quantity of carbonaceous matter, is full of a multitude of grains of such a deep green colour that they appear black; but they afford a green colour when crushed, and are, like those of the chalk, insoluble in nitric acid; above is a rock which is calcareous, granular, micaceous, sandy, of a whitish grey colour, and altogether resembling the craie tufau. It contains the remains of undeterminable shells.

These shells generally occur as casts, or rather as relievo

casts in the cavities of the shells, which have been almost always destroyed. These relievos are moreover ill shaped, entangled within each other, or joined together; yet they are sufficiently well preserved to determine with certainty the genera and species contained in the following list;

Organic remains of the upper and uncovered beds of the rocks and mountains of Fis, Sales, &c. forming part of the Buet chain in the Savoy Alps.

Nautilus			This is known to be a genus of shells, the greater part of the fossil species of which belong to the chalk.*
Scaphites obliquus	Sow.		In the chalk of Rouen.
Ammonites varians	Sow.		In the chalk of Rouen.
—— inflatus	Sow.		In the chalk (greensand) of Rouen, and in that of the Perte du Rhône.
—— Deluci	A. Br.	(pl. vi. fig. 4.)	In the chalk of the Perte du Rhône.
—— clavatus	Deluc.	(pl. vi. fig. 14.)	Coll. of Deluc.
—— Beudanti	A. Br.	(pl. vii. fig. 2.)	In the chalk (greensand) of the Perte du Rhône.
—— Selliguinus	A. Br.	(pl. vii. fig. 1.)	
Hamites virgulatus	A. Ar.	(pl. vii. fig. 6.)	
—— funatus	A. Br.	(pl. vii. fig. 7.)	
Turrilites Bergeri	A. Br.	(pl. vii. fig. 4.)	
——? Babeli	A. Br.	(pl. ix. fig. 16.)	
Trochus			An undeterminable cast but altogether resembling that which has been named Tr. Gurgitis.
Cassis avellana	A. Br.	(pl. vi. fig. 10.)	From the chalk (greensand?) of Rouen and the Perte du Rhône.
Ampullaria			Interior cast.

* Out of the nineteen species of nautili mentioned by Parkinson in his table of British fossil shells, (Introduction to the study of Fossil organic remains, (1822,) p. 231 & 232,) only five are from the green sand and chalk, so that it by no means appears that the greater part of the fossil species are found in the chalk, even when united to the greensand. (Trans.)

Cerithium...............................	Two species. They are broken, but are easily determined to be true cerithia; one of them so much resembles cerithium mutabile of Beauchamp, near Paris, that I cannot, at present, see any difference between them.
Inoceramus concentricus Park. (pl. vi. fig. 11.).	From Folkstone, Rouen, and the Perte du Rhône.
—— sulcatus ....Park. (pl. vi. fig. 12.).	From Folkstone and the Perte du Rhône.
Cytherea ?......... Cardium ?......... Pectunculus ?...... Arca..............	Casts absolutely undeterminable, even for the genus, except those of the arca, which very much resemble the interior casts of arca Noe.
Terebratula ornithocephala.* ..Sow.........	(Coll. of Deluc.) From the mountain of the Reposoir.
—— plicatilis............... Sow.........	Altogether resembling that found in the chalk.
—— obliqua ? .... Sow..........	
Echinus..................................	Oval, resembling mamillaris, but are much smaller.
Spatangus Cor anguinum.................	From all the chalk formations.
Nucleolites ? Rotula ..A. Br. (pl. ix. fig. 13.)	The echinites are very abundant, but all in a bad state.
—— castanea........A. Br. (pl. ix. fig. 14.)	
Galerites ? depressus..Lam. (pl. ix. fig. 14.) †	

* There appears to be some error here in the text, for three terebratulæ are given in the list. Parkinson, in his table of British fossil shells above referred to, gives, out of thirty-one species of fossil terebratulæ, six as found in the green sand, viz.: T. biplicata, T. intermedia, T. dimidiata, T ovata, T. pectita, and T. Lyra. The terebratulæ mentioned as occurring in the chalk and chalk marl, are, T. biplicata, T. subrotunda, T. carnea, T. semiglobosa, T. subundata, T. plicatilis, T. octoplicata, and T. obliqua. To this list should be added some others enumerated by Mr. Mantell, in his Geology of Sussex, viz. T. sulcata, T. Martini, T. striatula, T. squamosa, and T. subplicata. (Trans.)

† In order to compare this and the other lists with the fossils contained in the chalk and green sand of England, consult Conybeare and Phillips's Outlines, &c. p. 73, 74, 75, 76, 129, 130,—Parkinson's Table of British fossil shells, (Introduction to the study of Organic remains p. 231, &c.), and Mantell's Geology of Sussex. (Trans.)

It will be seen from this list, that the shelly deposite on the summit of the Montagne des Fis contains a great number of shells which almost exclusively belong to the lower chalk (green sand) formation. There are neither belemnites nor terebratulæ, because these shells, without being excluded from the green sand (craie chloritée,) are rarely met with in it.

The shells contained in this deposite so much resemble those of the green sand, that it was sufficient for me to name them. It will also be remarked how much the ammonites differ from those found in the body of the mountain.

I must confess, that notwithstanding the care I took to procure all the shells of this bed that I could discover at Servoz, among the guides at Chamouni, and at Geneva; notwithstanding those that have been presented to me by Messrs. Berger, Lainé, Soret, Selligue, and Beudant, or that have been sent me by M. Deluc; I must confess, I say, that this list is very incomplete; but it is sufficiently extensive to shew us the proportion in which the green sand shells occur in the Montagne des Fis, compared with that of the shells found in the same place, and which have not yet been met with in the green sand.

These zoological characters and analogies lead us to conclude that certain rocks of the Perte du Rhône and the summits of the Buet chain should be referred to the green sand formation, notwithstanding the mineralogical differences at first sight presented by the rocks composing these formations, and those which compose the generally admitted chalk formations. These differences are however greatly diminished by the green grains these rocks contain, and by the greyish and granular rock which covers them, so that the first circumstance, apparently so minute, finishes the analogies I have presented, and offers a new and remarkable application of what I have stated at the commencement of this notice, of the constancy of geological phenomena throughout all the known points of the earth's surface.

EXPLANATION OF THE PLATES.

Plate 8. *Fossil organic remains of the marl beds of the compact Jura limestone.*

Fig. 1. Strombus Pelagi. A. Br.

A and C. The same individual seen on different sides. B. Another imperfect individual, furnished with tubercles on the edge of the lip; they appear to indicate the bases of spines terminating this lip in the adult individuals.

The dotted lines represent those parts that are supposed to be wanting: it bears some resemblance to Strombus Gallus. (Coll. of Deluc.)

Fig. 2. Strombus Oceani. A. Br.

A. An individual apparently adult and complete, B. A younger individual.

Fig. 3. Strombus Ponti. A. Br.

A. An imperfect individual, but in good condition. B. An individual which appears imperfect, but which is worn.

These strombi are interior casts. The shell does not remain on any one. I have not found them either figured, described, or even noticed in any general work that I have been able to consult.

Fig. 4. A. B. Lutraria? Jurassi. A. Br.

The genus can only be presumed from the form, the hinge not having been as yet observed. This species does not appear to me to have been described or at least distinguished. It bears some resemblance to the mactra gibbosa of Sowerby, which is a Lutraria.

Fig. 5. A. B. Donacites Saussuri. A. Br.

I presume that this is a donax, but I am far from being certain. The form does not sufficiently shew it. The generic appellation should only be considered as provisional. (Coll. of Deluc.)

Fig. 6. A. B. Donacites Alduini. A. Br.

The figure is made from a tolerably entire and well preserved individual from Writhenterton in England, and which I consider perfectly resembles the shells collected at Cap la Hêve by M. Audouin, and by other naturalists in the upper marls of the Jura. It has much more the form of a donax than the preceding species, and even resembles the Venus meroe of Linn. I believe that this shell has been figured by Knorr among the (musculites) t. II. pl. B, 11, 6**, fig. 5.

Fig. 7. A. B. C. Cardium Protei. A. Br. Seen on three sides.

Though rather common in the upper marls of the Jura formation, I do not know it any where exactly figured or systematically named. The individual figured is from Cap la Hêve.

Fig. 8. A. B. C. Hemicardium tuberculatum. A. Br.

It slightly resembles the shell figured by Sowerby, pl. 143, by the name of Cardita tuberculata, and those figured by Knorr, tome 11. 1, tab. B. 1, a, fig. 2. 4; yet there is a difference. M. de Schlotheim first cited these shells of Knorr in the seventh year of the Taschenbuch, &c. of Leonhardt, by the name of bucardites reticulatus, and he afterwards cites them in his Petrefacten Kunde, &c. (published in 1820) p. 209, by the name of bucardites hemicardius. It will be easily seen why I have not admitted these specific names; the individual figure was sent me by M. Risso, it comes from the Cap St. Hospice, near Nice.

Fig. 6. A. B. C. Spatangus oblongus. Deluc.

M. Deluc sent me this name with this species, which occurs in the upper marls of the Jura. I conceive that the spatangus which I collected in the same marls, behind the town of Neufchatel, does not sensibly differ from it. It also appears to possess analogies with that mentioned by M. de Schlotheim in the Taschenbuch, &c. and the Petrefacten Kunde, by the name of echinites quaternatus.

Plate 9. *View and Section of the Montagne des Fis.*

Fig. 1. View taken from the Servoz foundry.

A. B. The part of the crest we attained. M. Kind of inaccessible col. *m.* the marteau. F. Behind this point is situated the bed containing the green sand fossils. S. Summit.

Fig. 2. Figurative section.

No proportions could be followed with regard to the thickness and extent of the beds, especially the collections of fallen masses, because the size of the figure would have been greatly augmented without any useful result.

Notices on the Hartz, by M. de Bonnard, Ingenieur en chef of the Royal Mining Corps.

(Annales des Mines, 1822.)

1. *Physical Sketch of the Hartz.*

If the Hartz is geologically considered, the formations of which it is composed may be regarded as belonging to the great schistose zone, which extends from W.S.W to E.N.E., from the north of France across the north of Germany; but the continuity of this zone suffers, at least on the surface, considerable interruption in Westphalia, and the Hartz forms in every respect a group of isolated mountains in the midst of rocks of more modern formation. This group is elongated from N.W. to S.E.; its greatest length is about 22 leagues, from Seesen to Friederich'srode, and its breadth 8 leagues, between Wernigerode and Walkenried. The Brocken, the highest summit of the Hartz and of the whole north of Germany, is, from the barometrical observations of M. de Villefosse, 1132 metres [3713 feet, 6 inches] above the level of the Baltic sea. The slope of the Hartz mountains, which is rapid on the north, is in general gentle towards the south, and especially the south-east.

Many small chains, or branches of mountains, generally more elevated than the others, quit the Brocken as a centre, and diverge in different directions through the Hartz. We should principally remark those two which divide the waters and which extend one to the N.E., the other to the south of the Brocken. On the east of this crest, the waters flow to the Elbe by several small rivers, the most considerable of which is the Bode; on the west, the waters flow towards the Weser, either to the south by the Söse and Oder which unite to join the Leyne, or to the north by the Innerst and Ocker. We should also remark the branch of the Lerchen Köpfe, which, commencing at the western side of the Brocken, and situated between the valley of the Ocker and that of the Radau, extends towards the N.W.; and especially that of the Bruchberg, whose commencement appears to be the same as the last, and which extending towards the S.W. separates the valleys of the Sieber and the Söse, and forms a nearly continuous ridge from the Brocken to the environs of Hertzberg, where it gradually lowers towards the plain. These small chains are generally elevated from 700 to 900 metres [2296 to 2952 feet] above the level of the sea.

Two principal and rather even platforms should also be noticed in the Hartz: the one on the west of the Brocken, comprises the towns of Clausthal and Zellerfeld with their environs; its general elevation is 580 metres [1902 feet, 6 inches]; the other, on the east, comprises the countries of Elbingerode and Hüttenrode, it is 50 metres [164 feet] less elevated. Cultivated fields occur on the latter, and rye and oats are reaped; potatoes were first cultivated on the former in 1806 round Clausthal. This last is still more remarkable for the fine meadows with which it is covered, furnishing an abundance of excellent hay, and mostly in two crops. The valleys which furrow these two platforms are not deep.

The remainder of the Hartz is in general uncultivated, except on the S.E. where the ground is much lower; the sides of the mountains are covered with fir-tree forests, or waste tracts, which are but the remains of ancient forests, destroyed, either by fires and storms, or by the ravages of an insect named borken köfer (dermestes typographus ?) or from the carelessness with which the wood was formerly cut in a climate whose severity renders numerous precautions necessary for the reproduction of forests. A few meadows are found at the bottom of the valleys, but they are of small breadth. The sides of the mountains, though steep, are uniform, and without remarkable escarpments in the greywacké of which the greatest part of the Hartz is formed: it is not the same with the granitic portions; the valleys of the Bode, the Ilse, and the Ocker offer both pictureseque and varied points of view, which recall to mind those of high mountain scenery. Without perceiving a very striking difference, we shall yet remark the depth and steep sides of the valleys of the Oder, the Sieber, and their confluents, i. e. in the environs of Andreasberg, a country placed between the two principal southern branches we have noticed, but formed of rocks of schist and quartz, different from greywacké, and considered to be of more ancient formation.

On the N. and N.E. the Hartz mountains are as it were suddenly cut off, and a few hills only are observed in the rich and extensive plains of Blankenburg, Halberstadt, Goslar, and Wolfenbüttel; on the S., where the slope is much less steep, a country of plains and low hills also occurs at the foot of the Hartz. On the S.E., on the contrary, the mountains gradually become lower, and the hills forming the Mansfeld country may be considered as constituting an appendage of the Hartz mountains.

II. *On the relative antiquity of the formations in the Hartz.*

When a group of mountains is formed, like the Hartz, partly of granite, partly of other crystalline and hard rocks and argillaceous or flinty slates which do not contain a trace of organic remains, partly of varieties of sandstone and pudding-stone known by the name of greywacké, alternating with a more or less micaceous clay slate (grauwacken schiefer), both containing the impressions of vegetables and shells, and partly of calcareous rocks, which appear entirely composed of madrepores and which occur united with the two last rocks, we are naturally led to consider the granite as constituting the nucleus of the whole; the crystalline greenstones (diabases), the diallage rocks, the *hornfels*, the quartz rocks, and the hard slates without fossils as immediately following the granite in the order of antiquity and as belonging with it to the primitive formations; lastly we suppose that the greywackés, the micaceous argillaceous slates and the limestones, whose place is generally well determined among the transition rocks, must rest on the other rocks.

Such in fact is the opinion long since adopted with respect to the rocks of the Hartz, and independently of general analogies which may lead us to conceive it, local observation affords a sufficient number of facts for its support. Quitting the Brocken, and proceeding towards the N. or N.E., that is in the space comprised between this mountain and the towns of Neustadt, Ilsenburg, and Wernigerode, we scarcely find any thing, besides the granite, but these crystalline and hard rocks which are regarded as primitive; and in many places we may conclude from their disposition that they immediately rest on the central granite. Proceeding southward from the Brocken, analogous rocks are met with in the environs of Andreasberg, with schists and flinty slates which do not contain a trace of organic remains, and they are sometimes seen to rest on granite; a similar superposition may elsewhere be concluded from the dip of the schistose beds. Proceeding still further south, we observe that the reputed primitive rocks are evidently covered by greywacké. Lastly, throughout the whole eastern parts of the Hartz, we remark a general dip of the greywacké beds to the S.E., leading us to consider it as resting on the granite of the Brocken, or on the other rocks previously observed.

The mineralogists who have inhabited, or until lately visited the Hartz, have conceived ideas slightly different from each other with respect to the antiquity of the *hornfels*, the quartz rock, and the hard schist of the environs of Andreasberg; but they are all agreed in recognising the primordial nature of the granite, and its anteriority with respect to all the other rocks, which they regard as above it. M. Hausmann, who in 1807 published a very instructive geological description of the Hartz,* has not only expressed the same opinion, but it appears from what Messrs Lamé and Clapeyron state, still preserves it, after subsequent observations undertaken for the sake of verifying it. Supported by this new and imposing sanction, this opinion would appear to merit entire confidence. Yet a very different opinion has for many years been propagated among German geologists, and it appears to me interesting to make it known. It has been very concisely exposed by M. de Raumer in 1811, in his *Fragmens géognostiques*; M. Schulze has adopted it in his memoir on the Hartz, inserted in the *Annuaire minéralogique* of M. Leonhard for 1815; since that time this opinion has been defended and opposed in many German works with which I am unacquainted.† I shall, consequently, in the following notices, principally rest on my own observations, by which I was, in 1806, led to conceive an analogous opinion; but I shall confine myself to the exposition of *doubts*; the extreme reserve we should always impose on ourselves when we wish to draw any conclusion from geological observations, is imperiously demanded, when this conclusion is opposed to an opinion supported by M. Hausmann.

It is more particularly when studying the western part of the Hartz, that we may be led to doubt the primordial nature of the granite of the Brocken. It should in the first place be remarked that throughout the Hartz, and especially throughout the greywacke formation, which constitutes at least three-fourths of the mass of the mountains, the beds are observed to have a general direction towards the E.N.E. and a general dip towards the S.S.E. with but a few local exceptions. It has been seen, in the preceding notice, that the Hartz mountains are elongated from north-west

* Geognostiche Skizze von Süd-Nieder-Sachsen, inserted in the 2d No. of the Nord-Deutsche Beyträge zur Berg und Hüttenkunde. Brunswick, 1807.

† In a very interesting memoir, inserted in the Annuaire minéralogique of M. Leonhard, 1821, and which only reached me after these notices were written, M. Germar opposes the idea of the comparatively modern nature of the Hartz granite, which he considers as of primordial formation.

to south-east: thus the general direction of the beds is perpendicular to that in which the group is elongated.—The Brocken is situated a little to the west of the line which would form the small axis of this kind of ellipse, and to the N.N.W. of the centre. Around the Brocken, the whole of the rocks regarded as primitive fill nearly a triangular space, which would have the northern limit of the Hartz for a base, from the *usine* of Ockerhütte (half a league to the E. of Goslar) to the town of Wernigerode, and whose summit would be placed a little to the S. of the town of Andreasberg. The same rocks and granite reappear in many places, on the S.E. of this triangular space, in the midst of formations recognised as transition; but throughout the west and north-west portions, no trace of these presumed primitive rocks is discovered, except it is on the prolongation of the Bruchberg branch on the west of Andreasberg, a little beyond the triangle. Now, from the general dip to the S.E. this western part is precisely that whose beds appear to dip beneath all the others, and in many points they seem to dip beneath the granitic rocks which are situated beyond them. I am aware of the errors we expose ourselves to when we wish to conclude a superposition solely from observations of this nature, and when no point of direct superposition has been proved; but it must be confessed that this general dip towards the S.E. observed throughout the eastern part of the Hartz, is one of the facts brought forward to support the antiquity of the crystalline rocks, relatively to the greywacke of this part of the group; and it is at least very remarkable, that on the western limit of these primitive rocks, we should no where observe the greywacke with a western dip, whilst in many places, on the contrary, it shews an opposite dip. In the eastern part, the granite occurs in many places in the midst of the recognised transition rocks, and there has no where been noticed in the beds of the latter, that variety of dip which would denote their mantling round the different granitic nuclei; the dip remains almost constantly the same.*

* M. Germar expresses an opinion in his memoir, that the schistose

The rocks of hard schist and flinty slate in the environs of Andreasberg, which are also considered as of primitive formation, certainly appear in some places to rest immediately on granite or on the *hornfels* covering the granite; but their general dip, as well as that which can be observed with respect to the quartz rock of the Bruchberg, is nearly that of the greywacke situated to the west of the Bruchberg, and which would consequently appear to be beneath them. M. Schultze notices similar rocks in the eastern part of the Hartz, which appeared to him to be of contemporaneous formation with the greywacke, and to present many passages into and mixtures with the latter. Many local observations tend equally to make us doubt the constant anteriority of the Hartz granite, relatively to the other rocks that have with it been regarded as primitive.

Ascending the valley of the Ilse, we find steep granite rocks one league from Ilsenburg, known by the name of Ilsensteinsklippe; situated opposite each other, on both sides of the valley, which becomes narrow as it approaches them, and expands above them, they appear to form part of a granite bed situated in the midst of different rocks which have less resisted the action of the waters. Lasius has long since shewn that these rocks exhibit very marked traces of a stratification parallel to the general stratification of the Hartz rocks. Ascending higher up the valley, schistose and quart-

beds of the north of Germany, and particularly those of the Hartz, having a general direction from W.S.W. to E.N.E. should not be deranged in their dip by the salient masses of the more ancient crystalline rocks they meet with; that thus they would not present round these primordial masses the variety of dip, which is erroneously required as a proof of their superposition, and that they might occur dipping towards the granite, on the west and north of the granitic mountains, as is seen in the Hartz, without our being able from that circumstance to deduce any probability of contemperaneous formation. The force of this reasoning appears to me much diminished from the examination of the Iberg mountain, where, as we shall presently see, the greywacke presents very different sides of the transition limestone; and it appears very difficult to conceive why the primordial masses have not produced on the transition slates as much effect as the calcareous masses that are scarcely anterior to these slates.

zose rocks are found with the same stratification; still higher up granite is met with.

At the rocks of Rosstrap, on the banks of the Bode, at the eastern extremity of the Hartz, I conceived that I also observed a marked stratification in the granite, and a general dip of the beds to the east. I also saw beds of quartz, schistose greenstone, and a species of mica slate, which appeared to me to alternate with the granitic beds. M. Schultze notices a general stratification, at least as apparent, in the numerous granitic rocks of the Hohneklippe, to the E. of the Brocken. The same person has observed a bed of granite included in rocks of flinty slate aud claystone *(argilolite)* at the foot of the Ramberg, a granite mountain situated one league to S.S.E. of Rosstrapp.

I have observed on the Adenberg, to the N. of the Hartz, at the extremity of the Lercheukӧpfe branch, and on the right bank of the Ocker, a bed of granite distinctly included between the beds of quartz rock and flinty slate of which the mountain is composed. M. de Raumer mentions granite as mixed with the *hornfels* on the Sandbrink, which forms part of the same chain. This mixture is again found, in small alternating beds, in the mountain of Rehberg, where however in the end the *hornfels* covers the granite. On the N.W. of the Brocken, in the mountains which border the Radau, and which are entirely formed of diallage rock (long known by the name of greenstone or primitive trap), many beds of granite have been observed included in that rock.*

* M. Germar describes the position of these granitic beds; they have, he says, a direction from S. to N. and dip towards the E. The diallage rock that contains them does not offer any positive stratification; yet the author considers these granites as evidently forming beds, and he derives, from their different position from that of the greywacke beds, another conclusion against the contemporaneous origin of the two formations. He also supports his opinion by the circumstance, that, throughout the Hartz, the granitic masses appear elongated from S. to N. and consequently in a transvere direction to that of the transition rocks. He considers that these granitic masses of the Hartz correspond with those of the same nature situated in the Thüringerwald, nearly under the same meridian.

Let it be recollected that many of the granitic rocks of the Hartz are true syenites, that some others are *protogynes*,* and that *protogynes* and syenites are often considered as belonging to the transition class. Let us also consider, with M. Raumer, that the environs of Dohna in Saxony, present above a schistose formation containing beds of greywacke, rocks analagous to those which accompany the Hartz granite, among others the *hornfels*, and lastly the granite itself. Let it be added that there are many among the mineralogists inhabiting the Hartz, who, while they believe in the primordial nature of the granite, do not consider the *hornfels*, the quartz rock of the Bruchberg, and the schists of Andreasberg, as of primitive formation.† Lastly, let it be recollected that M. Freisleben thought he found fragments of gneiss in the granite of the Brocken.‡

The general direction of the beds, from W.S.W. to E.N.E. is now considered to be, if not altogether constant, at least very general in the ancient rocks of the north of France, Belgium, and the north of Germany. I have elsewhere ‖

I do not conceive that the granite, constituting the mass of the Brocken and its environs, can be regarded as much more elongated from N. to S. than from E. to W.; but it is true that at the eastern extremity of the Hartz, the granite of Rosstrapp, of the Ramberg and Auerberg (especially the first and third) are situated nearly south from each other. I do not know if this circumstance is sufficient for M. Germar's opinion: as to his ideas with regard to the correspondence of the granitic rocks of the Hartz and Thüringerwald, he only presents them as a sketch not founded on precise observations. Not being acquainted with the Thüringerwald, I can neither support nor oppose this resemblance, which is very different from what I am about to state; I shall only observe that it appears to me opposed to the received ideas of the general disposition in Europe, either of rocks of the same nature, of the great lines of elevated countries, of low countries, of lakes, or inland seas.

* A name given by M. Jurine to the granitic rock of Mont Blanc, described in a previous part of this section. (Trans.)

† M. Germar observed at the old mine of Glückauf, near Andreasberg, a well characterized greywacke slate. He also considers that all the schistose rocks of this canton are not of primordial formation, but contemporaneous with the greywacke of the remainder of the Hartz.

‡ See Géognosie de Reuss, t. 2, p. 211.

‖ Sketch of the coal measures of the north of France, &c. Journal des Mines, No. 156, p. 418, &c.

noticed an example of this fact in the great zone of the Belgic coal measures, and I have stated that if the line drawn from Liège to Valenciennes, a line very exactly indicating this general direction, is prolonged to the westward, it would in Normandy pass very near the Litry coal measures. Further still, at the extremity of Britanny, the coal measures of Quimper are situated a little below the same line. I have moreover mentioned, at least as an extraordinary circumstance, the position of the two coal basins of Sarrebrück and Montrelais (the first we are acquainted with to the south of the preceding line), on another line nearly parallel to the first. I shall now notice another fact of the same nature, which belongs to the object of the present notice, and which appears to me equally worthy of attention.

A granitic formation is known to exist in the department of La Manche, on the north of the peninsula of the Cotentin; it is known that this formation contains granites, syenites, and *protogynes*, and that the observations of Messrs. Brongniart and Omalius d'Halloy, lead us to regard the whole as of contemporaneous formation with the quartz rocks and schists of the Cotentin and Britanny, some of which contain organic remains. It is also known that this granitic formation constitutes on the E. the most northern capes of the Cotentin; whilst on the western coast it is found a little further south, so as to indicate a direction from W.S.W. to E.N.E., parallel to that of the schistose rocks of the same country. Now if a straight line is drawn on a map from the granitic mountains of the Hartz to the granitic capes on the E. of Cherbourg, this line when prolonged would traverse the Cotentin in the direction of the granitic band of that country, and it moreover would very nearly be parallel to the line above mentioned as the general direction of the Belgian coal measures. It should here be remarked that the Cotentin and the Hartz are, in this general direction, the two most northern points in which granite appears on the surface in all that part of Europe which is situated to the south of the Channel and the Baltic Sea, and that this rock is no where found in the interval which separates them;

but that the granitic porphyry knolls, which are isolated in the midst of the schistose rocks of Hainault, occur at a very short distance from the line which would join these two granitic points.*

Without pretending to draw from this fact, considered alone, conclusions which would at least be very hazardous, I conceive that it may be regarded as supporting the facts which lead us to doubt the primordial nature of the granite and the other crystalline rocks of the Hartz, and as making it presumable that this country offers another example of the return of the most crystalline ancient rocks in the midst of transition formations.

The transition rocks of the Hartz, at least those generally recognised as such, considered alone, present among themselves relations conformable to those elsewhere observed in similar formations. The general dip of the beds to the S.S.E. leads us to suppose that the more ancient parts of the schist and greywacke formation should occur at the northern limit of the group, and there, in fact, we observe either a characterised clay slate, long regarded as primitive, and which very rarely contains some scarcely determinable remains of marine animals, or a greywacke of such a fine and close grain, that its arenaceous structure is with difficulty distinguishable. Advancing towards the S. and S.E., i.e. on beds resting on the preceding, the schists become less fissile, more dull, more micaceous, more shelly, sometimes contain small rounded grains, and thus pass into greywacke; this last becomes visibly arenaceous, and the rolled fragments it contains, increase in bulk so as to acquire the size of the head; it moreover contains numerous remains of plants, the surface of which is penetrated by anthracite, and entire sandy and often slightly calcareous beds (psammites sableux of M. Brongniart), filled with the shells usual in this formation.

The greywacke of the Hartz contains, as Messrs. Lamé

* See Essay on the Geology of the North of France, by M. Omalius d'Halloy. Journal des Mines, No. 142, p. 304, &c.

and Clapeyron have observed, numerous subordinate beds of alum slate (ampelite), whetstone slate, flinty slate, and hornblende rocks of different kinds, as also numerous beds of iron ore. Clay-stone porphyry (argilophyres) and other porphyries are also known there, the relative position of which does not always appear well determined; yet some certainly appear as beds included in the greywacke. We should remark, among these last, one which is analogous to the vert antique porphyry, and which occurs between Rübeland and Elbingerode, and another porphyritic rock, whose paste is a mixture of compact whitish felspar and quartz, and contains crystals of roseate felspar and hornblende. I have observed this last rock forming the roof of the iron ore beds of the environs of Elbingerode. The greywacke also contains beds of limestone, many of which are altogether analogous to the *marbres campans*.

The transition limestone of the Hartz presents two considerable independent masses. One situated at the western extremity of the group, near the small town of Grund, constitutes the Iberg and some neighbouring mountains. The greywacke rests upon it, as may be seen by pits driven through the greywacke to reach the iron ore worked in the limestone. It should be remarked that the stratification of the greywacke, round the limestone nucleus, is *mantle shaped*, since the beds of greywacke and slate dip to the N.W., on the western sides of the Iberg and the Bauerberg, a striking exception to the general dip of the Hartz. The other limestone mass, situated in the environs of Rübeland, on the E. of the Brocken, also occurs in some places beneath the slates; but it appears to form a very thick bed or parallel mass, included between beds of greywacke. The limestone of this formation appears almost exclusively composed of madrepores, which only become visible from the decomposition of the rock. Both masses contain caverns: two of those in the environs of Rübeland are known by the names of Baumanns höhle and Biels höhle; those of the Iberg, which are less celebrated, but much more interesting, are partly filled with iron ore.

In the plains situated on the N. and N. E. of the Hartz, we almost immediately find the modern secondary formations; on the W. and S. we observe more or less of the ancient secondary formations; on the S.E., the general series of these formations and those which follow them are developed in the Mansfeld country, so well described by M. Freiesleben. It will be seen that the greater or less continuity or interruption in the general series of formations, corresponds here very exactly with the less or greater steepness of the slopes.

On the calcareo-trappean formations of the southern foot of the Lombard Alps. By* ALEXANDRE BRONGNIART, *Member of the Royal Academy of Sciences, &c.*

(Annales des Mines, 1822.)

THE author describes by this name the formations situated at the southern foot of the Lombard Alps, which are composed of calcareous, trappean, amygdaloïdal, and basaltic rocks alternating together, formations previously described by Arduino, and more especially by Fortis, and which are for the most part situated in the Vicentin.

He does not commence by a detailed account of these formations; he confines himself to mentioning the rocks, their distribution, and the other circumstances which are necessary to prove the truth of the resemblances he considers he is able to establish between these formations and those with which he compares them. M. Brongniart has visited five principal places, the characteristic features of which he notices as follows.

1. The Val Nera. We here see a remarkable alternation of horizontal limestone beds and a small grained trappean conglomerate, which has been named *tuf*; but as this name very ill applies to rocks which bear no real analogy to each

* The memoir of which the following extract appeared in the Annales des Mines, was published separately and more at large this year (1823) with the title of Memoire sur les terrains de sediment superieurs calcareo-trappéens du Vicentin, &c. [Trans.]

other, the author names the conglomerate, trappean brecciola (brecciole trappéenne). This brecciola, which is not a basalt, nor even a compact lava, alternates with a limestone containing nummulites and some fossil shells, the analogy of which to those of the calcaire grossier of Paris is remarked by M. Brongniart. Higher up, towards the commencement of the valley, the basalt appears and seems to rise from the midst even of the Brecciola.

This formation of brecciola and *upper sediment limestone*, commonly named tertiary limestone, appears to fill the bottom of a great valley, formed, previously to this deposite, in a compact and much more ancient limestone, occurring in an oblique and unconformable stratification with the brecciola formation. The author refers this limestone to the Jura limestone; as also a great part of that which occurs at the foot of the Alps in the same geological situation.

2. The Val Ronca, celebrated for the great abundance of shells found there, in general presents the same structure; but the alternation is less regular, the brecciola is in a thicker mass, and the basalt more abundant; the yellowish limestone, which even mineralogically resembles the calcaire grossier of the environs of Paris, is filled with a multitude of nummulites. These fossil shells, which have rendered the place so celebrated, are scattered in the brecciola beneath the limestone beds. The author gives a very detailed enumeration of these shells, with descriptions and very exact figures of all those which he has not found described, or which have not been so well figured as to be recognised. The shells, to the number of more than eighty species, principally described and figured from specimens and the information of M. Maraschini, of Schio, are all so similar, even as to species, to those of the calcaire grossier of the environs of Paris, that we may, in many cases, regard them only as simple varieties; more than twenty are even analogous to the species found in the Paris basin, and the author has then contented himself by mentioning them by the names given them, either by M. de Lamarck, or other conchologists. Among the analogous species are the following:

Turritella incisa, very close to the elongata of Sowerby. Turr: imbricataria, of *Lam.* Ampullaria depressa, *Lam.* Amp: spirata. Melania costellata, *Lam.* Nerita conoidea, *Lam.* Natica cepacea, *Lam.* Natica epiglottina, *Lam.* Conus deperditus, *Brocc*: Ancilla callosa, *Defr*: Voluta crenulata, *Lam*: Marginella eburnea, *Lam*: Murex tricarinatus, *Lam*: Cerithium sulcatum. Cerithium plicatum, and more than a dozen other species of cerithium. Fusus intortus. Fusus noæ. Fusus subcarinatus. Fusus carinatus. Fusus polygonus. Pleurotoma clavicularis, &c. The descriptions and figures can alone give a certain and useful idea of the others.*

3. Montecchio-Maggiore. The trappean formation is here so predominant, and of such a crystalline structure in some of its parts, that it is more difficult to recognise here, at first sight, the same origin and association of rocks than in the preceding places; yet, if the limestone is not found in alternating strata, it occurs in inclined superincumbent beds, and we recognise the epoch of this formation in the fossil shells which are disseminated, not in the amygdaloïdal nodules, for they contain none, but in the brecciola that unites them. These shells are of neighbouring species, and sometimes absolutely the same species as that of the two places above noticed, and consequently of a contemporaneous

* The author, in his separately published memoir gives a long list of the fossil shells found in the calcareo-trappean rocks of the Vicentin. The list is too long for insertion here, I have therefore extracted from it the following catalogue of the genera with the number of their species. (Trans.)

Nummullites, 1 species. Bulla, 1. Helix, 1. Turbo, 2. Monodonta, 1. Turritella, 4. Trochus, 2. Solarium, 1. Ampullaria, 6. Melania, 3 Nerita, 3. Natica, 2. Conus, 2. Cypræa, 2. Terebellum, 1. Voluta, 3. Marginella, 2. Nassa, 1. Cassis, 3. Murex, 2. Terebra, 1. Cerithium, 14. Fusus, 5. Pleurotoma, 1. Pteroceras, 1. Strombus, 1. Rostellaria, 2. Hipponix, 1. Chama, 1. Spondylus, 1. Ostrea, 1. Pecten, 2. Arca, 1. Mytilus, 3. Lucina, 2. Cardita, 1. Cardium, 1. Corbis, 2. Venus?, 2. Venericardia, 2. Mactra?, 2. Cypricardia, 1. Psammobia, 1. Cassidulus, 1. Nucleolites, 1. Astræa, 1. Turbinolia, 2.

epoch. The author remarks the presence of lignite fragments and sulphate of strontian as points of resemblance between the formation at Montecchio and that of the two following places.

Monte-Viale. We here see in a very clear manner the alternation of the brecciola and limestone; but in certain parts of this hill, the two rocks are, it may be said, placed separately, and the basalt forms a distinct group on the confines of the hill. Fewer shells are found at Monte Viale than at Ronca, but those observed are of the same epoch. The sulphate of strontian which sometimes fills the cavities of these shells, is a more striking fact here than at Montecchio; and the lignite, which occurs here in thin beds, contains the remains of fish. These circumstances lead to the determination of the epoch of the fifth and most celebrated place.

5. Monte-Bolca. The trappean and calcareous rocks still alternate here in an evident manner; but this alternation takes place in such considerable masses, that it sometimes escapes our attention; the limestone predominates; it appears removed from the calcaire grossier by its compact texture and fissile structure, but these are only mineralogical differences, which should yield to the geological relations derived from the union of all the other circumstances, and especially the presence of organic remains, such as the nummulites, some shells of the genus avicula, the fish that before appeared at Monte-Viale, the various plants, principally terrestrial, and all dicotiledons, the subordinate lignites, and the absence of any organic remain that would indicate a more ancient formation.

It results from these comparative descriptions, rendered more clear by sections of the rocks and figures of the fossils:

1st. That these five places, at no great distance certainly from each other, belong to the same epoch of formation, and that other places should be added to them, such as Monte Glosso, to the W. of Bassano, which the author has also visited, as also the Val-Sangonini in the Bragonza, Castel

Gomberto in the Valdagno, and many points of the Monte Berici, which the author has not visited.

2dly. That all these formations are analogous, in all their important characters, to the upper sediment formations commonly called tertiary; and consequently to the marine formations above the chalk of the Paris basin. But as two epochs of these formations have been recognised, one beneath the gypsum and the other above it, M. Brongniart has endeavoured to determine to which of the two they should be referred. He remarks that the presence of shells, which much more resemble those of the calcaire grossier below the gypsum than those of the upper marine formation; that the presence of certain species, such as nummulites, the Nerita conoidea, the Caryophillites, &c. which have only yet been found in this lower formation; that of the lignites, the fish, and the chlorite or green earth, all of which appear to belong to it; that the absence of sandstone and mica, or at least the rarity of this substance, so abundant in the upper formation, present an union of characters which would induce us to refer the calcareo-trappean formation of the Vincentin to the calcaire grossier beneath the gypsum of the Paris basin, which consequently places its formation at an epoch anterior to that in which the rocks (also named tertiary), were deposited that constitute the subapennine hills, so well described by M. Brocchi.

The presence of basalts and trappean rocks seems, in the first instance, to be peculiar to the tertiary rocks of the Vicentin, for this rock is not known in the formations of the environs of Paris; but, besides considering it as the product of local phenomena and peculiar to the north of Italy, M. Brongniart conceives that a resemblance may be found, (though very distant certainly), between the grains of green earth disseminated through the lower strata of the calcaire grossier and the decomposed and even loosely aggregated trappean rocks, which generally constitute the predominant substance of the brecciola, a substance also mixed with the limestone; so that this rock appears to differ from the chlorite limestone (calcaire chloritée) of the lower strata of the calcaire grossier in the environs of Paris, only because the limestone is there more abundant than the green earth, whilst in the Vicentin the trappean rock predominates.

We cannot follow the author through all the developements he gives these objects of comparison, nor his citations of the nataralists who have more or less approached this result; but we cannot avoid citing Mr. Buckland with him, as having, during the tour he made in Italy nearly at the same time as M. Brongniart, conceived a similar opinion as to the epoch of formation of these rocks.

In a second memoir,* which M. Brongniart has not yet read to the Academy, he refers some other places he has observed or become acquainted with to the same formation, i. e. to the lower marine formation or calcaire grossier of the environs of Paris; such are among others:

1st. The high hill of the Superga, on the E. of Turin, principally composed of calcareous marl and calcareo-serpentine brecciola, containing shells for the most part analagous to the species of Bordeaux, Chaumont, and some other places which decidedly belong to the lower formations of the upper sediment (tertiary) rocks.

2dly. The summit of the Diablerets chain, above Bex, in the Valais. This rock differs from that of Paris, by its position, by its elevation of at least 2400 metres [7,874 feet] above the level of the sea, by the black colour and hardness of its calcareous and bituminous beds, but which may apparently be referred to this formation from the nature of the organic remains it contains, which are cerithia, ampullariæ, a cardium, very near the ciliare of M. Brocchi, if it is not the same, the melania costellata, or a very near species, a hemicardium, which is analogous to the retusum or medium, &c.†

* This forms the second part of the author's recently published memoirs on the Vicentin rocks. (Trans.)

† We must not confound this with another rock, which appears to resemble it in its colour, position, &c.; but which essentially differs from it in its shells; and which forms part of the mountains of Sales, Varens, &c. on the S.W. of the Buet. The author, in a memoir inserted in

3dly. He also refers to this formation, but still more doubtfully, circumstances not having permitted him to observe this rock in place and in detail, the granulated green rock, sometimes mentioned by the name of green sand (grès vert), which occurs on the summit of the high alpine limestone mountains at the opening of the valley of Glaris, near Nefels, and perhaps in many other places,* rocks which contain the remains of shells that generally resemble those of the upper sediment (tertiary) rocks, and especially a great quantity of nummulites, which, as is known, tolerably well characterize this formation, yet without exclusively belonging to it.†

another part of this selection, has described the latter rock among those he has referred to the green sand.

* M. Brongniart adds the following note, in his separately published memoir on the Vicentin, to his notice of the Glaris rocks. (Trans.)

" M. A. Boué has observed a similar rock, which he names grès vert, and which he refers to the green sand (glauconie crayeuse), in two places of the first line of the N.W. slope of the Alps, on the side of Bavaria, 1st. near Sonthofen, 2ndly near Trauenstein. These rocks rise from 300 to 1000 metres [984 to 3280 feet] above the valleys; their nearly vertical beds incline to the S.; they rest on smoke-grey compact limestone (zechstein), and even here and there on *more recent* deposites. The formation is principally composed of quartzose sandstone, chlorite or ferruginous sandstone, of compact glauconie (green sand), of brownish or reddish limestones, with disseminated nodules of granular hydrate of iron; these calcareous rocks are full of nummulites. We also observe in this formation, and especially in the ferruginous beds, Belemnites, Ammonites, Ananchites, Gryphites, Pectens, Sharks teeth, &c.

" Now if all these shells are associated in the same beds, as the note M. Boué addressed to me seems to indicate, these formations resembling that of Glaris in the green arenaceous rock, and the nummulites would certainly belong to the green sand, and these observations would make it presumable that the grès vert of Glaris, Sarnen, &c. belongs to it also." Memoire, &c. p. 50 & 51.

† The author also refers some rocks in the environs of Mayence to the calcaire grossier. In which opinion he is supported by M. Steininger who however does not consider the fresh water formation as alternating with it, but as resting upon it. (Trans.)

Notice on the Magnesite of the Paris Basin, and of the position of this rock in other places. By* ALEX. BRONGNIART, *Member of the Royal Academy of Sciences, &c.*

Read to the Royal Academy of Sciences, April 1, 1822.

(Annales des Mines, 1822.)

The distribution of the rocks and minerals entering into the composition of the crust of the globe, may be regarded in different points of view, and the different kinds of relations subsisting between these bodies successively examined.

Sometimes we take a formation composed of different kinds of rocks, whose epoch of formation is well determined in one place, and we follow it in other parts of the globe, to see if it preserves the same position, and to study the mineralogical modifications it experiences: this point of view is *principally* geological and *secondarily* mineralogical. Sometimes we study a simple or mixed rock, of a certain nature, and following it in different places or in the different formations in which it occurs, we examine at what epochs it has been deposited on the surface of the globe, what are the minerals and rocks with which it is associated, and what peculiarities it presents in each of these epochs. This point of view is *principally* mineralogical and *secondarily* geological:

* This paper is perhaps more mineralogical than geological, yet as it involves geological considerations I conceived that it would not be out of place in this selection. (Trans.)

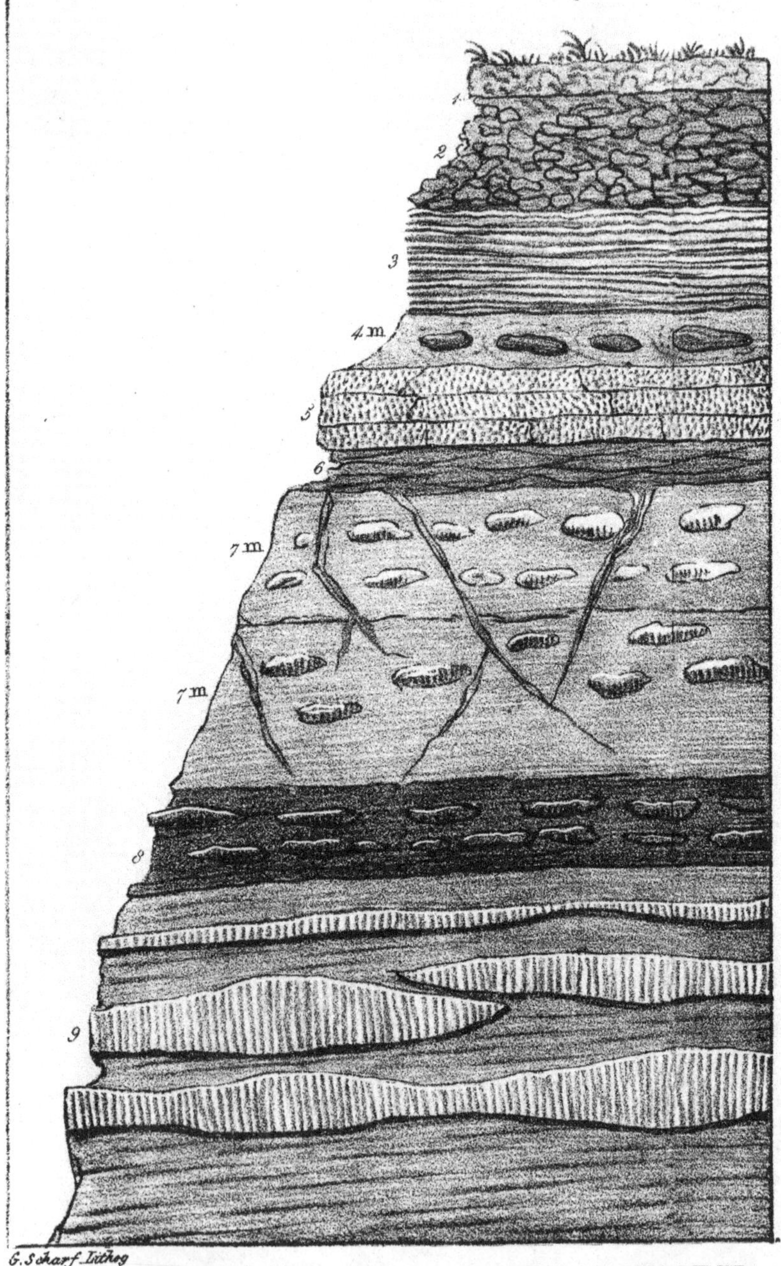

MAGNESITE OF VALLECAS, NEAR MADRID.

London. Pub: by

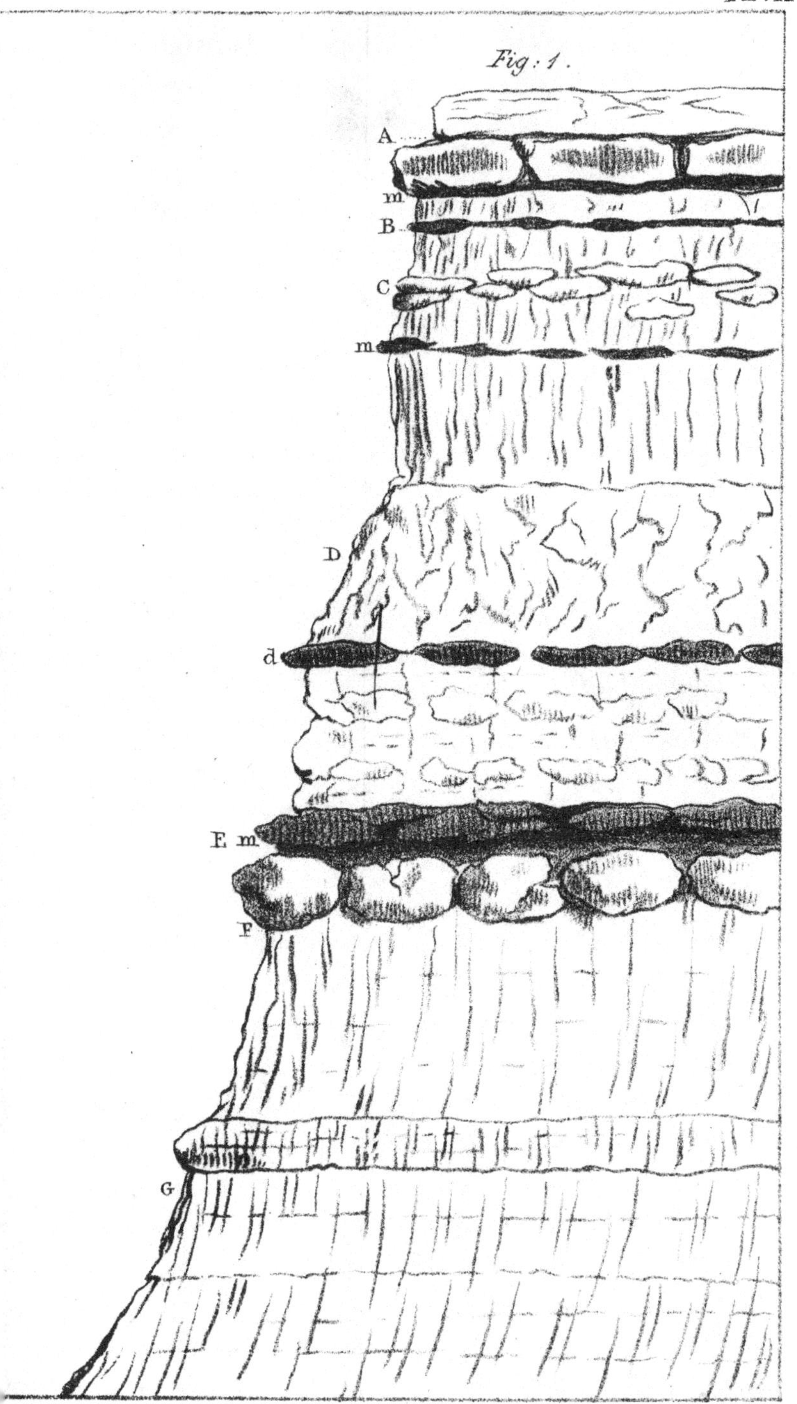

MAGNESITE AT COULOMMIER.

Phillips. 1824.

it is as productive as the first in general results, and consequently as proper as it to discover the laws which have presided at the structure of the earth, and at the formation of the minerals that enter into its composition.

It is under this last point of view that I shall consider the mineral which I have mentioned by the name Magnesite.

The following are the minerals to which I give this name. I distinguish them in two principal series, which may one day be separated into two species when we shall have observed sufficiently essential characters to establish this distinction.

1. Plastic magnesite (magnésite plastique), composed of magnesia, silex, and water, without carbonic acid.

I here comprise the magnesite, so improperly named écume de Mer, that of the environs of Madrid, that of the environs of Paris, that of Salinelle, department of the Gard, &c.

Serpentine might, from its composition, almost be referred to this species; but it is distinguished from it by its mineralogical characters.

2. Effervescent magnesite (magnésite effervescente), essentially composed of magnesia and carbonic acid, sometimes associated with very variable proportions of silex and water.

We may refer to this division the magnesite of Hroubschitz in Moravia: those of Piedmont, of the Isle of Elba, of Baumgarten in Silesia, of Styria, &c.

Having made known, as far as it appears necessary, the minerals I include under this name, I shall now describe the position of the magnesite of the Paris basin, and present the union of a few facts and observations in order to complete the geognostic history of these minerals, the principal object of this notice.

Parisian Magnesite.

I first observed the presence of magnesite in rather extensive beds at Coulommiers, twelve leagues to the E. of Paris, and afterwards quite close to the latter town: I shall describe this variety and the circumstances of its position with some detail, as I shall afterwards employ it as a type of comparison with the same mineral, found in other positions and in other places.*

The magnesite of Coulommiers, in the purest specimens, for it is often mixed with other things, possesses the following characters:

Its masses are soft, smooth to the touch without being unctuous; its powder is rather hard.

It easily absorbs water and swells out considerably, becomes slightly translucent, and forms a short soft paste, resembling jelly.

It does not effervesce with acids.

Exposed to the action of a porcelain furnace (at 140° of Wedgewood), it hardens, exfoliates a little, but does not suffer any other alteration; it does not shew the slightest trace of fusion, either in its thin pieces or on the surface; it however becomes rough to the touch, and hard enough to scratch steel.

M. Berthier has analysed this magnesite, chosen from the purest masses, and has found the following ingredients:

Magnesia	24
Silex	54
Water	20
Alumine	01,4
	99,4

The magnesite of Coulommiers occurs in masses, which, by their schistose structure and thinness, shew they belong to thin beds.

Its colour is whitish, most commonly pale grey; it has often a roseate tint, but it loses that and its grey colour in the fire. It is associated with brownish and reddish chert (silex corné) of a very scaly fracture; it is intimately united with it, and penetrates into all its cavities, and even into its

* I am indebted to M. Merimée for the knowledge of this magnesite. He was struck with the soapy unctuosity of a stone which he found at Coulommiers, and having brought it to me, he put me in the way of discovering this mineral in the Paris basin.

mass; it is also very frequently associated with marly limestone, and then effervesces and becomes partly fusible.

This magnesite occurs in thin beds, interposed between beds of marly limestone and calcareous marl, near Coulommiers, on the right of the road, entering the town on the Paris side, in a small hill having a north and south direction, and which having been cut to form a canal, exposes its interior structure and the following series of rocks, beginning with the uppermost. (pl. 10, fig. 1.)

1. A bed A, composed of siliceous limestone, the middle of which is of white and cellular chert (silex corné), and the compact limestone mass filled by small shells scarcely determinable, and by larger shells, such as Limneus longiscatus, cyclostoma mumia, &c.

2. This bed rests on a bed B of very irregular thickness, of a greyish fissile earth, resembling clayey marl, and which has been recognised to be an impure magnesite (m.), i. e. mixed with calcareous marl.

3. Then follows a bed of soft and friable calcareous marl, containing another small bed of magnesite (m).

4. A bed of calcareous marl without silex, beneath which is another small bed of brown impure magnesite.

5. A thick bed of white calcareous marl D, subdivided into many strata by marl beds, and by a bed of zoned chert (silex corné zonaire), almost jaspic, without either shells or magnesite.

6. A bed E about two decimetres thick, composed of brown chert (silex corné) in irregular nodules, but principally flattened. These are the nodules that are enveloped and even penetrated by the Parisian-magnesite of an isabella roseate grey colour (m). It is sometimes very pure, does not effervesce with acids, and is absolutely infusible in the heat of a porcelain furnace. It is sometimes slightly translucent.

7. These cherts (silex) are placed on a bed F of hard calcareous marl in nearly round nodules, and containing cyclostoma mumia.

8. Beneath is a thick bed G of white calcareous marl, friable or only splintery, and containing neither chert (silex) nor shells.

The total thickness of the beds composing this hill is nine metres [about 29 feet].

As this succession of beds and rocks is isolated, as no other formation is seen above it, and as we do not know that on which it rests, we can at most suspect its position by a comparison of these rocks with those that resemble them in the Paris basin, but this is a presumption difficult to prove without the presence of the organic remains found in it; now this character, which is so useful in establishing analogies between formations far distant from each other, possesses all its value when it is required to determine the position of one formation with respect to the others in the same basin: it may then be here employed with perfect safety, and geologists who admit these rules of determination, and who have seen the cyclostoma mumia and Limneus longiscatus cited, have immediately recognised the position of the formation containing the magnesite of Coulommiers. These shells are not marine, one of them is evidently a fresh water shell, consequently the magnesite belongs to a fresh water formation, and the two species of shells I have just mentioned, having as yet been only found in the middle fresh water formation, in that situated between the two marine formations of the Paris basin, we should refer the magnesite of Coulommiers to that fresh water formation; it forms part, as we have elsewhere * shewn, of that which we have named siliceous limestone. The hard calcareous marls, and the silex that accompanies the magnesite, remind us of the siliceous and calcareous characters of this deposite, and complete all the analogies.

The magnesite having shewn itself in a very distinct manner, both as to its purity and quantity in the siliceous limestone of Coulommiers, the rules of geology teach us that we should find it elsewhere, by searching for it in this for-

* Description géologique des environs de Paris, 1822, p. 38, and 203.

mation; this has in fact happened. Proceeding towards Paris, and at about two leagues from Coulommiers, we observe near Crécy the same rock with the same mineralogical circumstances; i. e. the limestone so compact that it resembles the fine compact limestone of the Jura, the chert (silex), the clayey marls, the magnesite, but less pure, and the same fresh water shells.

The short distance of these two places rendered these resemblances very presumable; but transporting ourselves to St. Ouen, close to Paris, on the bank of the Seine and at the foot of Montmartre, we find the magnesite in a formation altogether similar to that of Coulommiers; the same limestone, the same chert (silex), the same shells occur there; the position of the rock beneath the gypsum is there well determined. The magnesite is however less pure here and less apparent; traces of it only occur; these traces had long since been observed. M. Armet had remarked the presence of magnesite in the marls of Montmartre; M. Bayen had observed, more than thirty years since, and had shewn me that the menilite contained it. Now this belongs to the fresh water formation beneath the gypsum; it is probable that we should find this mineral either in minute quantities, or in small masses, in all the siliceous limestone rocks of this same formation, such as those of Champigny, Orleans, Septeuil, &c. I have recognised it in a greyish clayey marl which accompanies a silex resinite of the environs of Mans, consequently at more than 40 leagues to the west of Paris, and 50 leagues from the first place in which I have mentioned it.

Geological circumstances of the magnesite of different places, compared with those of the Parisian magnesite.

We shall find this rock still further distant, in a basin separated from ours not only by a distance of more than 120 leagues, but by chains of mountains whose structure and nature are altogether foreign to those which surround our basin; now, it is remarkable, that we find the magnesite with all the circumstances which accompany it in that part of the Paris basin where it is most pure.

Magnesite has long since been observed at Salinelle near Sommières, in the department of the Gard, between Alais and Montpellier; but its position has only been determined a few years since, by the description M. Marcel de Serre has published of it.

It is therefore solely to the remarkable analogy of this position with that of Coulommiers that I wish to call the attention of naturalists. The magnesite of Salinelle is schistose like that of Coulommiers; it possesses the same colour, approaching grey with a roseate tint, with the same tenacity; it absorbs water in the same manner; it is composed of the same ingredients, i. e. 20 parts of magnesia instead of 24, 51 of silex instead of 54, and 22 of water instead of 20. It will be acknowledged that it is difficult to meet with more resemblance between uncrystallized minerals, which occur at more than 100 leagues from each other, and if the mineralogical species cannot be here determined by the form, it is sufficiently so by the composition: the analogies drawn from its associated minerals, and its position are the same; it is mixed with nodules of chert (silex corné) which resemble our menilite; it is accompanied and covered by marly limestone containing fresh water shells, consequently it belongs, with that of the Paris basin, to a calcareo-siliceous fresh water formation.

But magnesite, i. e. this stone essentially composed of magnesia, silex, and water, occurs in many other places dispersed over the surface of Europe, and consequently placed at great distances from each other. Sometimes we are acquainted with its mode of occurrence, and then we know that it is very different from that I have above described; sometimes we are ignorant of it, or at least we do but presume it: but in all these places and in all these positions we shall see the magnesite to occur accompanied by the same mineralogical characters and the same *geological cir-*

cumstances (circonstances geologiques); * a consideration that must not be confounded with the geological position (gisement.)

The magnesite of Vallecas near Madrid is already known, for in 1807 I described, in my Traité de Mineralógie (t. 2. p. 492.), its nature and properties, from the information obtained by the specimens received from Messrs. Sureda, Dumeril, and Mieg, and of its position from the same specimens, and the information of M. Link, who took it for a kind of clay; a very excusable error at that time. M. de Rivero has however studied the same places, and has sent me an ideal section of this rock, which I have given, pl. 10, fig. 2. and a detailed description which I shall transcribe almost literally.

" The village of Vallecas is two leagues to the south of Madrid; it is situated lower than the latter town; an isolated hill, named the hill of Vallecas, occurs near the village: before we reach the top of this hill, we meet with small hillocks and excavations which arise from the workings of the magnesite; the tour of this hill may be made in 20 minutes. From observing the locality, an idea is conceived of a gypsum basin on which the magnesian rock rests."

" If we observe the structure of the hill, we observe, commencing at the lowest part (pl. 10. fig. 2.), N° 9, gypsum with clay, which belongs to the saliferous formation † of Villarubia: this gypsum extends from the walls of Madrid to the junction of the river Javama with the Manzanares; it is very distinctly seen near the hermitage of Notre Dame de la Torré, 150 metres [492 feet] to the west of the hill of Vallecas and near the canal of Madrid; there then follows a bed of reddish clay (N° 8.) with nodules of flint (silex pyromaque.) Though the magnesite has not been observed immediately on the clay, yet M. de Rivero conceives that it

* I have literally translated M. Brongniart's expression, though I should not have used it myself in the same sense; M. Brongniart seems only to imply that it is constantly associated with certain minerals, without any reference whatever to its geological or relative position. (Trans.)

† New red or saliferous sandstone. (Trans.)

rests upon it, because ascending towards the hill, the magnesite is found to follow; and the flint nodules are the same as those of the magnesite. The magnesite (N° 7. m.) occurs in very thick beds, coating flints which are disseminated through the beds: these beds are cleft, and in the clefts we find asbestus (asbeste papyriforme), on which crystals of carbonate of lime are observed; they are also seen on the magnesite. This same deposit re-appears close to Madrid, it may be observed as we leave the *barrière* by the Portello; the flint is there disseminated in the same manner. M. de Rivero has also met with it on the banks of the river Manzanares, opposite the king's villa; it has also been found at Cabanas, nine leagues to the north of Madrid: the author, not having visited this last place, is unable to describe its situation. A thin bed of greenish clay (N° 6.) containing very little magnesite is observed above the magnesite at Vallecas; then follows a reddish common opal (silex resinite) in beds of variable thickness, very fragile, presenting a crust of manganese on some parts of its surface; this opal is worked for gun flints. A very soft and nearly earthy magnesite (N° 4.) is found above this fragile opal."

"The different beds above noticed by M. de Rivero, occur in the hill of Vallecas. The top of this hill constitutes a platform, on which are found many flints, and pieces of opal, with crystals of carbonate of iron; crystals of pseudomorphous quartz have moreover been observed, and have been taken for opal crystals."

"Shells have never been met with in this formation. The beds, represented in the ideal section, fig. 2, by the Nos. 1, 2, and 3, appear on the banks on the Manzanares, as we quit the gate leading to the Escurial.

The author has above stated that magnesite is met with on the banks of the river, and if we ascend towards the town, we find beds of greenish and reddish clays (No. 3), of which bricks are made, and above these clays an alluvial formation (No. 2), composed of fine grained sand, and lastly vegetable earth, (No. 1) on the surface."

Thus the magnesite of Vallecas and Cabanas, near Madrid, possesses the same tenacity, the same hardness, the same lightness, the same superficial roseate tint, as those of Coulommiers and Salinelle. It is equally composed of 23 parts of magnesia, 53 of silex, and 20 of water; it is accompanied, like ours, by chert (flint?), which also passes into its mass, by common opal (silex resinite), by chalcedony, by crystallized quartz, and calcareous spar altogether resembling those of our siliceous limestone. It affords, certainly, no organic remains; but we know that these remains are rare in the siliceous limestone of the Paris basin, of which our magnesite forms a part; lastly if it appears to differ by its position on a saliferous gypsum, much more ancient than our gypsum, and calcaire grossier, it is not covered by any rock which appears more ancient than the latter, and it is like them in horizontal beds.

If from Spain we transport ourselves to Italy, to the foot of the Piedmontese Alps, we shall find, at a short distance from Turin, the serpentine hills of Castellamonte and Baldissero, traversed in every direction by veins of magnesite which is tenacious yet plastic, light, and with that roseate superficial tint which we have noticed in the preceding magnesites. Its principal or fundamental and characteristic composition appears to be still the same, i. e. of magnesia, silex, and water. Here however we have carbonic acid, which seems to indicate a different chemical species; but its *geological circumstances* are still the same. I have already noticed them in my memoir on the geological position of the serpentines, where I have given a figurative section of them.

The mineral no longer occurs in horizontal beds, or nodules interposed in the beds, but in numerous veins, uniting in every direction in the midst of the serpentine; chert, common opal, and jasper, presenting many varieties of texture and colours, are constantly and intimately united with it, as at Coulommiers and Salinelle. They have been formed even in the midst of the magnesite. This circumstance of geological association is then remarkably constant, even when the geological position has no longer the same character, and it is here very different. It appears to me well established, that this magnesite belongs to the serpentine formation of the Apennines, consequently to ancient rocks, nearly of the transition epoch.

There are other examples of magnesites, but the circumstances of their geological position are less well known; yet both what is known, and their composition, still very well agree with what we have stated of the preceding.

Thus the plastic magnesite of Asia minor, known by the name of Ecume de Mer, has all the exterior characters of that of Piedmont, and even that of Coulommiers, with a composition that very slightly differs; it has, like it, the roseate superficial tint which also occurs in the magnesite of Houbricht in Moravia. But in this, the carbonic acid, which is in some quantity, seems to establish a mineralogical difference, the importance of which is not yet well appreciated; the presence of silex nodules which pass into the mass, reminds us of an analogy in the *geological circumstances*, which is rather remarkable.

CONCLUSIONS.

We shall confine ourselves to these examples: they are sufficient to prove the relations of formation which we wish to establish between the magnesite of the Paris basin and those we have just mentioned. The magnesite in all, whether it be or be not combined with carbonic acid, contains water and silex: this last substance does not occur only in chemical combination with the magnesia, it also forms isolated masses, and whatever the mineralogical differences may be that these varieties of quartz present, not only is its presence all that is necessary to establish the geological resemblances which we desire should be remarked; but it may be said that these varieties follow without interruption from the oldest to the newest magnesites, as the following table will shew:

Parisian Magnesite	Crystallized quartz Chert Several varieties of opal (silex resinite)
Magnesite of Salinelle..	Chert
Magnesite of Madrid .	Crystallized quartz Chert (silex corné) Chalcedony Several varieties of opal (silex resinite)
Magnesite of Moravia..	Chalcedony White and green opal (silex resinite)
Magnesite of Piedmont	Chert Chalcedony Varieties of opal (silex resinite) Jasper

Before geology had acquired in principles and facts the precision to which it has now arrived, the presence of magnesite in the Paris basin had no other result than that of adding a mineral species to the list of those contained in our country; but this fact now possesses another interest: it has served to unite observations which were, it may be said, isolated. It informs us that the magnesite beds were deposited on the surface of the globe at very different epochs, for some (those of Piedmont) belong to the most ancient sediment rocks, and others (those of Salinelle and Coulommiers) to the newest sediment (tertiary) rocks; and yet we see these deposites accompanied by nearly the same *geological circumstances.* Such a remarkable constancy in the association of silex and magnesia, two bodies between which there is no chemical analogy, will fix the attention of geologists, and may perhaps contribute to shew us the origin of these deposites, as the thermal springs of Italy depositing travertine have pointed out that of the freshwater limestone. It is still apparently from the bosom of the earth that the liquid arose which deposited these rocks; for we find in certain thermal waters traces of all the ingredients of their composition: the mass of water is at present immense in comparison with the matters held in solution; but these matters exist in it: they are deposited, as M. Berthier has observed at the waters of Vichy, St. Nectaire, &c.* not only separately but nearly in the same order as the

* Annales de Chimie et de Physique, t. xix. p. 134.

calcareous and magnesian formations. The first deposites, those which are nearest the spring, this able chemist tells us, are also those most charged with peroxide of iron and silex; the limestone, still ferruginous, then follows, and is the more pure and more separated from these two substances, the more distant it is from the point where the spring rises from earth; the carbonate of magnesia is the last deposited.

Without wishing to establish any real resemblance between this succession and that of our rocks; without wishing to represent that these rocks, certain beds of which shew too clearly the characters of mechanical aggregation for them to have been formed by solution, have been deposited by the mineral waters of the ancient world, we cannot avoid remarking that commencing with the chalk, we find a series of rocks, the nature and succession of which are nearly the same as those which M. Berthier has observed in the deposites from mineral waters. Thus, 1st, a new formation, i.e. a new emission of dissolved matter would appear to commence above the chalk, at first depositing silex and iron, represented, one by the beds of sand and sandstone, and the other by the iron ore found so abundantly in the deposites of lignites and plastic clays which cover the chalk; 2dly, the more or less compact limestone, accompanied by iron and silex in the lower beds, and by silex in the upper beds; 3dly, the magnesite also accompanied by silex, which still occurs in the lower gypsum beds; this silex is partly soluble in alkaline liquids, like that of the calcareous deposites of certain mineral waters; 4thly, the gypsum, the most soluble substance of all those we have named, and which should be the last deposited.

We do not pretend to draw any other conclusion from these different resemblances; but it appeared to us right to *hazard* them, if it were only to engage the attention of chemists and geologists.

*Observations on a sketch of a Geological Map of France, the Pays-Bas, and neighbouring countries; by J. J. d'Omalius d'Halloy.**

(Annales des Mines, 1822.)

When Baron Coquebert de Montbret was charged with the direction of the Statistics of France, he conceived the project of a general description of this vast state, which should be established on less variable bases than political and administrative divisions, and which should avoid the repetitions which the particular description of each of these artificial divisions requires; he, in consequence, was desirous of forming divisions into physical regions; but he felt that hydrographical basins, though invariable, were not more productive of general results than political divisions; he on the contrary considered that the only divisions fit to attain the proposed end, were those derived from the nature of the ground. The productions of a country in fact depend on this circumstance, and notwithstanding the modifications that may arise from manners, governments, and other causes, the inhabitants are generally in constant dependance on the productions of their soil.

* This memoir was composed at the end of 1813; but the author, called for some time to duties which did not permit him to occupy himself with natural sciences, has been obliged to delay its publication until the present time; but he hopes that this circumstance will plead his excuse for not placing his work on a level with the progress that geology has made within the last ten years.

A sketch of this nature, which I published in 1808, furnished M. de Monbret with the idea of engaging me to construct a map which should represent the mineral masses of different kinds which cover the surface of France; he promised at the same time to direct me by his advice, to place at my disposal the numerous materials he had assembled in the course of his studies, and to facilitate the acquisition of new information, both by his correspondence from the Bureau de Statistique, and his personal acquaintance with the most able mineralogists, especially the Engineers of the Mining Corps of France. I eagerly embraced a proposition which so completely entered into my pursuits, and which would place at my disposal an union of means superior perhaps to those enjoyed by any naturalist; but I soon perceived the almost insurmountable difficulties of this work. I observed that independently of those which result from the nature of the work, the information with which I was surrounded left immense deficiencies, that many observations which went back to epochs, anterior to the progress geology has latterly made, had become useless, and that far from being able to supply them by my own labours, it would require the whole lives of many laborious men to unite the necessary materials. These obstacles would have made me renounce the enterprise if I had not felt that this kind of work had better be ill done than not at all, since it may in some manner be said, that error in this case leads us to truth. I also thought that the kind of sacrifice I should make of my self-love for the advantage of science, would entitle me to indulgence. I have then constantly devoted to this work the little time that duties foreign to science left at my disposal, and I have made several tours which, though too rapid to procure me a true acquaintance with the countries I passed through, yet furnished me with the means of arranging the observations of others on the same plan.

Two principal points of view seem equally to lead to the division of a country into physical regions determined by the nature of the soil: in one it is considered geologically, i. e. according to the epoch of formation, in the other with respect to its mineralogical and chemical nature. We might at first sight imagine that the latter way was the best to attain the end proposed, as it seems to have most connexion with the action peculiar earths exercise on vegetation; but, on the other hand, the different states of aggregation of the substances composing the rock, the physical position of the soil, and other circumstances belonging to the epochs of formations, often exercise an equally marked influence. Thus the pasturages of the Pennine Alps, the heaths (garrigues) of Languedoc, and the fields of la Beauce, present very great differences in their aspect and productions, though the soil is calcareous in all. It will moreover easily be perceived that the geological mode of consideration is much more advantageous for the progress of science, that it offers much more interest, and that the power it allows of uniting, according to circumstances, many systems into one group, permits us to dispense with the detailed observation which would be required, in the other case, for the frequent changes of the predominant substances, in a formation of the same epoch.

It may be supposed that, this principle being once adopted, it would be sufficient to take as guides the divisions established in geological works, and to mark on the map the places where the different formations determined by systematic authors occur; but experience soon proved to me that I could not follow this mode, though so simple in appearance, for admitting all the subdivisions established in geological treatises, we should infinitely multiply the obstacles resulting from the want of observations sufficient for such a large extent of country, and from the difficulty of referring to common terms the different systems that exist in distant countries. It sometimes happens that rocks which are greatly developed in some countries fine off to such a point in others, that it is no longer possible to mark their existence on a general map. It has therefore been necessary to sacrifice to uniformity the very natural desire of presenting all the details we possessed on certain countries; it was also necessary to form a system which, while it left out a great part of the divisions established by authors, should yet preserve the most essential sections, and should agree with the developement of the different rocks in the countries I wished to represent: it will easily be understood that if some rocks fine off so that their existence cannot be separately marked on the map, and that these rocks possess general relations which distinguish them from other groups, it would be better to represent them by a common sign, rather than to undertake a distinction that could only be adopted in a few instances.

Guided by these considerations, and after numerous trials, I have been led to the system which has served as a basis to the map of which I hazard the publication.* I am far from considering this system to be without objections, perfection will never be attained in a first sketch, if perfection could be found in the works of men.

I shall not now undertake a particular examination of the different groups of rocks traced on the map; but I shall shew the principles that have guided me in the formation of these groups.

The old division of rocks into *primitive* and *secondary*, i. e. anterior and posterior to the existence of organized beings, can no longer accord with the intimate union that has been remarked between some primitive rocks, and beds containing organic remains; the celebrated school of Freyberg introduced an intermediate class in which to place these last beds. Since that time, new observations have proved that these intermediate rocks, instead of being constantly posterior to all rocks that had the general characters assigned to primitive formations, occur included between crystallized rocks, which do not contain any organic remains.

* The work which I have undertaken conjointly with M. Coquebert de Monbret, containing many more details than could be represented in the small map joined to this memoir, we conceived it might be useful to publish it on a larger scale; we have in consequence made it the subject of another and more extensive map, on which M. de Monbret has moreover added various agricultural information, such as the limits of the countries in which the vine, olive, orange, &c. are cultivated.

We may deduce two important conclusions from these latter facts: the first is, that nature has been able to produce similar rocks at different epochs, and that consequently mineralogical characters are insufficient to determine geological divisions. The second is, that rocks until now considered as primitive may be posterior to those which contain organic remains: so that, in the actual state of our knowledge, it becomes very difficult to distinguish the true primitive rocks from the transition, and every great division departing from this principle is of difficult application. It should not therefore be surprising if I propose to unite these rocks into one great class, which I shall name *primordial rocks* (terrains primordiaux), which has already been employed to indicate a less exclusive property than that attributed to the word *primitive.*

These rocks possess a very important character, which is that their beds occur in inclined, disturbed, and even vertical positions. The most ancient secondary rocks certainly also present circumstances of this nature,* but less generally and not in so decided a manner.

I was desirous of tracing on the map the principal systems that are distinguished in the primordial rocks; but after having successively reduced the number of these divisions, I found myself obliged to renounce the project; for independently of the rocks presenting the union or rather the confusion of many formations, I must confess that I at present find the geological relations of all these divisions to be in the greatest uncertainty.

It must in fact be confessed, that our means of judging of the relative age of rocks are in the end reduced to the superposition of beds. All the other characters we employ for these determinations are but analogies drawn from the observation of places where the superposition is evident, and

* And also the newer secondary rocks, as for example the vertical beds of plastic and London clay in the Isle of Wight, (the latter being the equivalent of the calcaire grossier of the Paris basin,) as also the nearly vertical chalk of the same place and Dorsetshire. (Trans.)

where it does not appear that the primitive disposition of the beds has been deranged; but can we place true confidence in the superpositions observed in the primordial rocks in which the beds are often vertical? It is an opinion very generally adopted, that these beds have originally had a horizontal or slightly inclided position, and that they owe their present situation to violent causes: now, a cause sufficiently violent to throw a bed into a vertical position, may also have given it an inclination in a contrary direction, and consequently have placed that beneath which was at first above; we may the more easily admit this mode of superposition, as the numerous accidents of the inclined beds shew that their disturbance has not been the result of one catastrophe alone.

I am fully aware that the partisans of the exclusive anteriority of crystallized rocks, may retort this argument, by attacking the conclusions which I draw from the positiou of some of these rocks above those containing organic remains; but if they grant this principle, the whole of their system will fall to the ground, since the series of formations they adopt is only founded on these superpositions; the character of the absence of organic remains is but a negative fact of little importance in this respect, since the study of secondary rocks has proved that the deposites of a siliceous nature and those abounding in crystallized portions, commonly contain few or no organic remains, as if the liquids from which these deposites were made had driven the animals away.

The common divisions of primordial rocks not being applicable to the map of the countries on which I was engaged, I had the idea of substituting another, founded on the circumstance that we might recognise three bands in these rocks, which are distinguishable for peculiar characters. One of these bands, situated in the north, and which comprises Britanny, and the countries between the Scheldt, the Weser, and the Hartz, is remarkable for the abundance of slates, and transition limestone found there.

Another, on the south, which comprises the Pyrenees,

the Montagne Noire, and the Alps, is distinguished by a great quantity of slaty rocks containing more or less talc.

We are lastly, in the intermediate space, struck with the abundance of granite in the primordial rocks of the centre of France, the Vosges, the Forêt Noire, &c.

Yet I renounced the desire I had of making these distinctions on the map, because in the present state of my knowledge, I could attach no geological consideration to it, and because these distinctions might arise solely from certain systems of rocks, common to the three bands, being more developed in one than in the other.

In fact, if I were desirous of finding the most ancient of these three modifications, I should direct my attention to the summits of the Alps and Pyrenees, which have long been regarded as the crests of the ancient world; but I see, that M. Charpentier,* considers the granite of the Pyrenees less ancient than that of Saxony; that M. von Buch † believes that the granite of St. Gothard rests on talcose and calcareous rocks; and I observe that the granitic rocks of Mont Blanc bear great analogies to those in the Tarentaise, which M. Brochant has shewn ‡ to belong to the transition series. If I afterwards descend into the centre of France, to examine a granite altogether similar to the rocks of Saxony, which are regarded as of the most ancient formation, I remark considerable resemblance to the rocks which, in Britanny and the Contentin, rest, as in Norway, on transition rocks, and an intimate connexion with the secondary rocks evidently posterior to the transition series.

On the other hand, the union into a single group of all the primordial rocks, i. e. of those commonly called primitive and intermediate, has presented the most fortunate harmony with the physical and economical considerations that may be deduced from the work which forms the subject of this memoir; for nearly, with the exceptions that result from the mineralogical nature of some particular rocks,

* Journal des Mines, t. xxxiii. p. 101.

† Leonhard's Taschenbuch, &c.; 6. Jahrgang, seite 335.

‡ Journal des Mines, t. xxiii. p. 322.

from the disposal and elevation of the soil, from the formation of the valleys, and the existence of some superficial deposites, all these rocks present the same aspect, the same tendency to be covered with heaths, pastures, or forests, and the same difficulty of producing wheat; characters which are found equally on the slates of the Ardenne, the granites of the Limousin, and the talcose rocks of the Alps.

The *secondary rocks* do not offer the same uncertainty as the primordial, the superpositions are here evident, and although a part of them have suffered the effects of disturbance, it is not in so violent or irregular a manner, and we cannot here suppose that a large mass has really been turned over; the very general presence of organic remains offers many means of comparison between distant countries, especially since the brilliant progress this study has made of late years. Yet, the distribution of the liquids into particular basins in which the newest formations have been deposited, the tendency that these rocks have to change their aspect or rather to develope one system at the expense of another, according to countries, leave many doubts to be cleared up.

I have considered myself able to divide these rocks into five groups, of which I shall sketch the general characters.

The first has for its type the rocks known in German geology by the name of todt-liegende * or red sandstone, to which I unite the *macigno* of the Tuscans,† and many coal measures, particularly those of the centre of France. ‡

* The new red sandstone conglomerate of Devonshire, &c. (Trans).

† A rock composed of quartz, clay, and limestone, is named macigno, in Tuscany; it is an argillaceous and calcariferous sandstone which, from the nature of its composition being very constantly the same, deserves a very particular distinction. This rock commonly contains other substances, and especially mica, which I do not consider as essential to its composition. It is very abundant in the Apennines, where it characterizes a particular formation.

‡ I do not here cite one of the most important coal measures of continental Europe, that which traverses the north of France and the south of the Pays Bas, because these coal measures bear so great an analogy to the primordial rocks in which they are included, that I have not yet entirely abandoned the idea I advanced in 1808, that these coal measures should also be considered as true transition rocks.

Addition by Translator.—It is much to be regretted that the author

These rocks are so intimately connected with the primordial rocks, that it is often very difficult to trace the line of demarcation, and it is very remarkable that this union takes place equally with all the systems of primordial rocks. It results from this circumstance, that, always partaking of the nature and even the colour of the primordial rocks, that they immediately follow, they present the greatest differences with each other. Thus for example, if they succeed red granite, they are also red, and contain much felspar; if they follow talcose rocks, they are greenish grey, and of a composition analogous to that of these rocks. It even appears that in some countries, and especially in some cantons of the Alps, this formation is represented by calcareous beds interposed between two formations also calcareous, the separation of which they mark, and from which they are only distinguished by slight differenees in texture and colour; hence there is considerable difficulty in distinguishing the secondary and transition rocks of this country.

Another character of the rocks forming the group that at present occupies us, is, that they contain a great quantity of conglomerate (clastoïde), and arenaceous rocks, i. e. composed of fragments or grains of greater or less size, and of a more or less different nature; yet this texture does not belong exclusively to this formation, it is on the contrary found in almost all formations; it may even be said that we rarely pass from one formation to another without observing some beds which have this peculiar texture, but at no epoch have these rocks been so abundant or so general as this. Does not this shew that this epoch is one of the most remarkable presented in the study of the globe, and consequently that the division above proposed is one of the most natural?

has placed the coal measures partly in the primordial rocks and partly in his first secondary group, as he thus makes two things of that which is essentially but one; the coal measures of the Pays Bas are precisely the same as the English, and the transition rocks (according to the author) with which they are connected are our carboniferous limestones and old red sandstone.

The rocks classed in this secondary group do not, by themselves, form extensive countries; but they often appear in those countries where the primordial rocks predominate, principally in those of granite. At other times they shew themselves in the lower parts of masses covered by newer rocks: so that, with respect to the space occupied on the map, I should have united it to other formations, if their geological importance and distinctive characters did not forbid such an association.

I unite many systems of rocks to form the second group, the most important are known by the name of Zechstein, or older Alpine limestone,* variegated sandstone,† muschelkalk, quadersandstein, aud jura limestone.‡

These rocks are often very distinct, but some possess common relations, which justify the approximation that their geographical situation in France has allowed me to make. They are so intimately united with those of the preceding group, that it is very difficult to establish the line of separation. I even consider that if my work had been more especially applied to Germany, instead of France, it would have been better to arrange the two first systems in the preceding group; for zechstein often occurs subordinate to the red sandstone, and the variegated sandstone constitutes regions which very much resemble those in which the red sandstone predominates.

But on the other hand, the zechstein occurs in the Alps, the Cevennes, and the Pyrenees, with characters that so much approximate it to muschelkalk and jura limestone, that it appeared to me preferable to group it provisionally with those rocks, the more so as the materials I possess on France, would not have afforded me the means of keeping up the distinction throughout the mass. I moreover consider that the zechstein is rare in the part of France to the north of the Alps and the Cevennes, that a trace of it only exists there,

* Magnesian limestone. (Trans.)

† New red or saliferous sandstone. (Trans.)

‡ Oolite formation.

and that it almost always passes into the coal measures or the marly rocks, which are probably the representatives of the variegated sandstone, a system which is also very slightly developed in these countries. It may be useful to remark here that the rocks placed in this group offer, in France, one remarkable circumstance; which is that the beds on the north of the Jura and the Cevennes are nearly horizontal, whilst those that occur in the Jura, the Alps, the Cevennes, and the Pyrenees, have constantly an inclination which may be expressed by the term arqûre (arched, saddle-shaped): this difference would appear to be independent of the epochs of formations: it may however possibly arise from the former constituting hills, and the latter elevated mountains.

The chalk formation, such as I have determined it in a preceding memoir, i. e. comprising the tuffas, sands, and marls, which occur beneath the true chalk, constitutes the third group.

It must be confessed that this formation, considered in a purely geological manner, is not of more importance than many of those which I have noticed in the preceding group; but I considered that it should be distinguished on the map, on account of the extent it occupies in France and the Pays Bas, where it forms gulfs in the midst of more ancient rocks, and is distinguished from the neighbouring countries by peculiar physical properties.

I unite in the fourth group all the rocks posterior to the chalk, whose aqueous origin is not doubted. These rocks, which were but little known a few years since, occur almost every where, and their history now forms one of the most important parts of geology. Their number and the differences they present, would demand a subdivision, if their frequent superpositions would not render these details impossible in a general map.

These rocks are distinguished moreover in the different countries where they exist by remarkable differences, arising from certain systems being more developed in one place than another; thus in the north of Germany and the Pays Bas, they form sandy plains, which mix with the sands of the chalk. The limestone beds predominate in the Paris basin. Lastly, in the plains watered by the Saône, the Rhone, the Po, the Aar, the Danube, &c., we are struck with the quantity of rounded fragments which are buried in the sandy and clayey deposites.

The volcanic rocks have too different an origin, and their existence belongs to phenomena which are too remarkable, not to be noticed on a geological map, however small the space occupied by them may be; but the establishment of this group would present many difficulties, if I knew not where to stop among those rocks which bear a greater or less resemblance to the products of existing volcanoes. But now that the researches of M. Beudant have thrown new light on this branch of geology, I considered that I might class the products of modern volcanoes with the two systems known by the names of trachytic and basaltic rocks.

It should be remarked that this group does not correspond, like those which precede, to a fixed epoch of formation; for while volcanic rocks are now forming, there are trachytes and basalts of more ancient formations than some rocks classed in the preceding groups.

The exposition of these divisions would certainly have made the imperfection of our geological nomenclature felt, if its defects were not generally admitted; yet we have not effectually endeavoured to correct it. It must in fact be confessed that it is a task that would require considerable knowledge in the person who should undertake it, I therefore do not pretend to do so, but as it has fallen to my lot to establish some new divisions, I considered that they should receive new names.

The following are the names that I propose to give the five groups into which I have divided the secondary rocks.

I shall name the first *penean* rocks (terrains pénéens), which is but the translation of todt-liegende, and which may besides be considered as expressing the circumstance that the most characteristic beds are *poor* in organic remains.

The second group will be named *ammonean* rocks (terrains ammonéens) because all the systems of which they are composed have been formed at an epoch in which the very remarkable animals named ammonites existed.

The third, corresponding with the chalk formation, will be called *cretaceous* rocks (terrain crétacé.)*

The name *mastozootic* (mastozootique) will remind us that among these rocks the bones of the mammiferous quadrupeds have been found, the study of which has given rise to the great work that has, it may be said, created geology among us.

The fifth group will be described by the name of *pyroïdal* (pyroïde), which without expressing any thing affirmative as to the manner in which these rocks have been formed, will shew that they all resemble those whose igneous origin is demonstrated.

It is admitted that the best method of representing the different formations on a geological map is by different colours, yet it must not be concealed that the existence of many rocks in a small space, and especially their successive superpositions, give rise to many difficulties; for it will be perceived that one very important formation in a country may be almost constantly covered by another rock, without shewing itself in a manner that would be represented on a horizontal plane but by a very small space, or even none at all. This difficulty may be avoided in special descriptions by sections, which shew the interior nature of the country; but it will be seen that this method cannot be used in a map of the size of that accompanying this memoir. It is therefore necessary to observe that the distinctions made on this map are far from constantly shewing the exact limits; but we should consider them as only shewing which is the most

* It should be observed that in a division less adapted to the physical geography of France, this small group might have been united to the preceding, and that in that case the denomination of *ammonean* rocks would have been better preserved, as ammonites still existed when the chalk was formed. This denomination and that of *penean* rocks might still be preserved when the zechstein and variegated sandstone pass into the first group; for these two systems are not commonly very rich in organic remains.

abundant rock in particular countries, without excluding the idea that this rock may be accompanied by other formations. Thus, for example, though the passage from the primordial to the secondary rocks rarely takes place without presenting *penean* deposites, the latter have only been represented where they were known alone to cover a surface so extensive as not to overcharge the map with details which would injure the general effect.

Neither has any attention been paid to the patches of *mastozootic* deposites which cover almost all other formations, nor those superficial deposites of loose matters which generally cover all those rocks on which vegetation has established itself.

The greater part of the observations which have served for the formation of that part of the map containing the South of France, going back to an epoch in which there was not a very clear idea formed of the divisions now established among limestone rocks, this part of the work greatly requires revision. That which regards the Alps, the Pyrenees, and the Cevennes, may also be considered as a sketch of the manner in which I regard the geological nature of these mountains, rather than an exact representation of the places on which the different rocks appear; for independently of our being in want of observations on this subject, it is known that mountainous countries rarely afford demarcations. It should in the last place be added that the north of Germany has been traced on the map from the observations published by German authors, and solely with the view of shewing the connection of the different masses of rocks.

On the Geology of the environs of St. Leger sur Dheune (Department of the Saone and Loire.) By M. Levallois.*

(Annales des Mines, 1822.)

THE village of St. Leger, situated on the little river Dheune, is on every side surrounded by hills, which are in general well cultivated and without cliffs or ravines; so that the gypsum quarries are almost the only places in which we can study the geological structure of the country.

The hill rising above the village on the East, is that which contains the gypsums. It is elongated from South to North, and is cut off on the South, at less than two myriametres [about 15 miles] from St. Leger, by a small valley, on the other side of which the coal measures appear, which afterwards predominate in the commune of St. Berain; it is terminated on the west by the Dheune, and apparently extends some distance to the North.

The gypsum formation principally occurs half way up. It is very well exposed in three of the quarries worked in open day. The following beds may be observed:

* This notice forms part of an account of the gypsum quarries of the same place. (Trans.)

Beneath about 2 metres [about 6 feet 7 in.] of vegetable soil:

1 metre [about 3 feet 3 inches] of a compact and hard stone, though decomposing in the air. This stone is white with a slight greenish tint: it effervesces but very slowly with nitric acid, leaving a siliceous residuum. It is a marly siliciferous limestone: the bed is divided into thin strata.

1 ditto slaty marls, red in the upper part, green in the lower.

1 ditto of a marly siliceous limestone, analogous to that of which we have just spoken; it is only more compact, with a more conchoidal fracture. It has sometimes a reddish tint: some pieces are incrusted with calcareous spar.

0,15 ditto of red and green marls.

1,50 ditto of compact reddish gypsum, slightly mixed with marl.

0,50 ditto of gypsum in small pieces as if squared, disposed in the plane of the beds and separated by marls.

1,50 ditto of gypsum slightly mixed with marls.

1,80 ditto of poor gypsum in nodules mixed with marl.

0,50 ditto of red marls mixed with a little gypsum, and principally containing fibrous gypsum in veins parallel to the beds.

3 ditto nearly forming the white gypsum bed named the *galerie*.

2,33 ditto forming the bed of gypsum named the *fond*.

They no longer work beneath this bed, but marls mixed with gypsum are still found. Still further down sand is met with.

In these three quarries, the beds are nearly in parallel stratification. They succeed each other precisely in the same order and with the same thickness. They are in general slightly inclined, and appear undulated: at least we are led to believe so as they are slightly arched in the three quarries. The dip is most generally towards the west.

The gypsum is generally compact, of small fine crystalline grains, of a white or rose colour. Lamellar and perfectly diaphanous gypsum is also found. Fibrous gypsum frequently marks the separation of the different strata.

The mass of rock is often pierced by a kind of funnel filled with earth, which sometimes vertically traverses the whole height. These funnels are very inconvenient, as they allow the rain water to filtrate through them.

Slips are also remarked, in consequence of which the beds have descended more than a metre. It will be observed that these slides have taken place on inclined planes of soft clay, which fill the clefts.

I have said that the gypsum formation principally occupies the middle of the slope of the hill. Ascending a little towards the summit, we soon observe a secondary limestone, which covers the gypsum formation. It is on the side of the great road, and not far from the village of Charcey. The limestone is blueish grey and compact; it contains a great abundance of gryphites; belemnites, ammonites, and pectens are also observed. It forms nearly horizontal beds, of slight thickness, separated by strata of earth which facilitate the extraction of the stone. This limestone is principally employed for making lime, and building. The beds, which are of a deep blueish grey colour, and contain few shells, are used as marble by the masons of Dijon.*

The foot of the hill, principally on the South and West sides, is composed of sandstone. These sandstones are yellowish, micaceous, fine grained, slightly schistose, friable, often reduced to the state of sand, and form beds of little thickness and inclination. Advancing towards the South, beyond the small valley which, on this side, limits the gypsum formation, these sandstones present slightly different characters, and contain beds of coal. They are then generally larger grained: lamellar portions of felspar are seen in

* It appears from the author's description that the gypsum and red marls of St. Leger form part of the new red or saliferous sandstone, that they are covered by lias, and rest, as will be afterwards seen, upon coal measures. (Trans.)

them: they alternate with schists more or less penetrated by carbonaceous matter. Among the rocks composing this formation, there is one, among others, which is very remarkable: it serves as a roof to a thin coal bed which occurs near the surface. It is a white and very compact rock, containing here and there only a few spots of mica. It might be considered as one of the extreme limits in the series of arenaceous rocks.

This coal formation extends from hence into the commune of St. Berain. Many pits are there opened. The coal obtained is not of good quality; but from the proximity of the Caual du Centre, all the large coal is of advantage to Châlons and Lyon.

The gypsum formation, in the environs of St. Leger, rests upon coal measures, and is covered by gryphite limestone.*

The same relative position of these same mineral masses is observed near Conches, 1 myriametre [about $7\frac{1}{2}$ miles] to the W.N.W. of St. Leger. In fact, after having continually travelled on these sandstones nearly to this town, where we deviate a little to the left towards Chalancey, we soon meet with the red and green marls resting on the sandstones, and above them the gryphite limestone, in which the iron ore occurs as a bed; it is composed of hydrate of iron in very small agglutinated grains, or a kind of ferruginous oolite.† The marls do not appear there to be accompanied by gypsum; at least the presence of this mineral has not as yet been ascertained.

It should be remarked, that, from the observations of Messrs. Lamé and Thirria, the gypsum formation existing at Decize, department of the Nievre, rests on the coal measures, and is covered by a gryphite limestone very analogous with that of St. Leger.

* Lias. (Trans.)

† This is probably the inferior oolite above the lias. (Trans.)

APPENDIX.

Description of the Muschelkalk and Quadersandstein. By A. von Humboldt.

(Extracted from his Essai sur les Gisement des Roches, &c.)

Muschelkalk.

A formation which varies but slightly, and whose vague denomination of *shelly limestone* has caused it to be confounded out of Germany, with the lower or upper strata of the oolite formation (with the lias or forest marble). It is well characterized by its more simple structure, by the prodigious quantity of shells, that are partly broken, which it contains, and by its position above the Nebra sandstone (new red sandstone), and beneath the quadersandstein which separates it from the oolite formation. It covers a vast portion of northern Germany (Hanover, Heinberg near Gottingen, Eichfeld, Cobourg, Westphalia, Pyrmont, and Bielfeld), where it is much thicker than the zechstein or Alpine limestone (magnesian limestone).

It extends in southern Germany throughout the platform between Hanau and Stutgard. In France, where, notwithstanding the labours of Omalius d'Halloy, the secondary formations beneath the chalk have been so long neglected,

Messrs. de Beaumont and Boué have recognised it round the Vosges chain. The muschelkalk generally possesses pale, whitish, grey or yellow tints: its fracture is compact and dull, but the mixture of small laminæ of calcareous spar, arising perhaps from fossil remains, sometimes renders it granular and brilliant. Many beds are marly, arenaceous, or passing into the oolite structure. (Seeberg near Gotha; Weper near Gottingen; Preussisch-Minden; Hildesheim. Chert (hornstein) passing into flint and jasper (Dransfeld, Kandern, Saarbrück), are either disseminated in nodules in the muschelkalk, or form beds of small continuity. The inferior strata of this formation alternate with the new red sandstone (between Bennstedt and Kelme), or insensibly pass into the sandstone, by becoming charged with sand, clay, and even (to the E. of Cobourg) with magnesia (magnesian beds of the muschelkalk).

Subordinate beds. The Marls and clays so frequent in the oolite formation, the new red sandstone, and the zechstein (magnesian limestone) are rare in the muschelkalk. In Germany, this rock contains hydrate of iron, a little fibrous gypsum (Sulzbourg near Naumbourg), and coal (lettenkohle of Voigt; at Mattstedt and Eckardsberg near Weimar) mixed with aluminous schist and carbonized fruits (coniferæ?). The nearer coal is found to the tertiary formations, the more do at least some of its strata approach the state of lignite and aluminous earth.

Fossils. From the researches of M. von Schlotheim, and rejecting the beds which do not belong to the muschelkalk, the fossils are: Chamites striatus, Belemnites paxillosus, Ammonites amalteus, A. nodosus, A. angulatus, A. papyraceus, Nautilites binodatus, Buccinites gregarius, Trochilites lævis, Turbinites cerithius, Myacites ventricosus, Pectinites reticulatus, Ostracites spondyloides, Terebratulites fragilis, T. vulgaris, Gryphites cymbium, G. suillus, Mytulites socialis, Pentacrinites vulgaris, Encrinites liliiformis, &c. Some isolated beds of the oolite formation perhaps contain more fossils than the muschelkalk; but in no secondary formation do organic remains so uniformly abound as in that which now occupies us. An immense quantity of shells, partly broken, and partly well preserved, but strongly adhering to the stone (entrochi, turbinites, strombites, mytulites) is accumulated in many strata from 20 to 25 millimetres (about 1 in.) thick, which occur in the muschelkalk. Many species occur united in families (belemnites, terebratulites, chamites). Between these very shelly strata are disseminated ammonites, turbinites, some terebratulites, with their nacreous shells, the Gryphæa cymbium, and superb pentacrinites. Corals, echinites, and pectinites are rare. From the abundance of entrochi in the muschelkalk, this formation has received the name of entrochite limestone (trochitenkalk) in some parts of Germany. As a bed of entrochi often also characterizes the zechstein, and separates it from the coal measures, this name may lead us to confound two very distinct formations. The denomination of gryphite limestone (calaire à gryphites of the zechstein and of the oolite formation) and all those which allude to fossils, without indicating the species, expose us to the same danger. It is stated that the muschelkalk contains the bones of large animals (oviparous? quadrupeds, Friesleben, T. 1, p. 74; T. iv, p. 24, 305) and birds (ornitholithes of the Heimberg; Blumenbach, *Naturgesch*; 3[te.] Aufl. p. 663); but these bones may belong, as also the teeth of fish, to the breccias and marls resting on the muschelkalk.

Messrs. Buckland and Conybeare, during their tour in Germany, considered the muschelkalk of Werner as identical with the lias. I am inclined to think that there is rather a parallelism than an identity of formation. The muschelkalk occupies the same place as the lias, it equally abounds in ammonites, terebratulæ, and encrinites; but the fossil species differ, and its structure is much more simple and uniform. The muschelkalk strata are not separated by the blue clays which abound in the lias. The middle strata of the latter possess a dull compact and even fracture, much more resembling the lithographic varieties of the oolite formation than the muschelkalk of Gottingen, Jena, and Eichsfeld. M. Boué has recongnised the muschelkalk in

France, in the platform of Burgundy, near Viteaux and Coussy-les-Forges, near Dax, in the commune of St. Pan de Lon, &c. I have not observed this formation in the equinoxial part of America.

Quadersandstein (Sandstone of Kônigstein).

A very distinct formation (Banks of the Elbe, above Dresden, between Pirna, Schandau, and Königstein; between Nuremberg and Weissenburg; Staffelstein in Franconia; Heuscheune, Adersbach; Teufelsmauer at the foot of the Hartz; valley of the Moselle and near Luxembourg; Vic in Lorraine; Nalzen in the Pays de Foy, and Navarreins, at the foot of the Pyrenees) characterized by M. Hausmann, and for a long time confounded either with the quartzose varieties of the new red sandstone, the sandstone of the plastic clay, or with the sandstone of Fontainebleau, above the calcaire grossier of Paris: it is the white sandstone of M. de Bonnard, and the third formation sandstone of M. d'Aubuisson. Preferring geographical names, I often call this formation the Königstein sandstone, the new red sandstone the Nebra sandstone, and the muschelkalk the Gottingen limestone.

The quadersandstein has a white, yellowish, or grey colour, with very fine grains, which are agglutinated together by a nearly invisible argillaceous or quartzose cement. Mica is not abundant in it, and is always silvery and disseminated in isolated plates. It neither contains the included oolite beds, nor the flattened lenticular masses of clay (thongallen) which characterise the new red sandstone. It is never schistose; but is divided into very thick beds, which are cut at a right angle by fissures, and of which some easily decompose into a very fine sand. It contains hydrate of iron (Metz) disposed in nodules. The organic remains disseminated in this formation present, according to Messrs. von Schlotheim, Haussmann, and Raumer, an extraordinary mixture of sea shells (very analogous to those of the muschelkalk) and dicotyledonous phytolithes. In it have been found mytulites, tellinites, pectinites, turritellæ, and ostreæ, (with cerithia, but no ammonites; Habelschwerd, Alt-Lomnitz in Silesia), and at the same time the wood of palms, the impressions of leaves belonging to the class of the dicotyledons, and small deposites of coal (Deister, and Wefersleben near Quedlinbourg), very well described by Messrs. Rettberg and Schulze, and passing into lignite.

M. von Raumer had observed that the quadersandstein is separated from the new red sandstone by the muschelkalk; it is placed between this limestone and the Jura limestone, and consequently beneath the oolite formations of England and the continent. In this position we cannot consider it, with M. Keferstein (see his Essay on the mineral geography of Germany, T. 1. p. 12. and 48.), as parallel to the molasse of Argovy (mergelsandstein), which represents the plastic clay beneath the chalk. The nature of the vegetable remains contained in the quadersandstein, and its resemblance to the plänerkalk which belongs to the chloritous and sandy strata of the chalk, have caused it to be regarded by many celebrated geologists as a formation posterior to the oolite formation: thus Messrs. Buckland, Conybeare, and Phillips place it between the chalk and the upper beds of the oolites. But, according to the observations of M. Boué and many other celebrated German geologists, the quadersandstein, sometimes alternating with marly and conglomerate beds, rests immediately on gneiss near Freyberg, on the coal measures in Silesia, and in Bohemia; on the new red sandstone near Nuremberg in Franconia; on the muschelalk between Hildesheim and Dickholzen near Helmstadt, and near Schweinfurt on the Mein. It is covered by the oolite formation, and alternates with marly beds of this limestone in Westphalia, between Osnabrück, Bielfeld, and Bückebourg.

On Muschelkalk and Quadersandstein.
By A. Boue'.

(Extracted from M. Boué's Memoir on Germany, inserted in the Journal de Physique for May, 1822.)

Second Secondary Limestone or Muschelkalk.

THE second secondary limestone or muschelkalk of the Germans is the least variable of all the German limestone deposites, and nearly always occurs in the same manner; from this great uniformity of character, and the extent of this formation it would appear surprising that so few foreign geologists should have a clear idea of it, or that they should not have recognised it in their own countries. But the explanation of this fact is simple as it respects England, the deposite not existing there, they have vainly endeavoured to recognise it. In France and Switzerland, this limestone being of little extent compared with the oolite formation, it has until now been considered, that the French muschelkalk formed only a part of the latter, whilst in the north of Germany, where the muschelkalk occupies, comparatively with the respective extent of both countries, nearly as much space as the oolite formation in France, it has been considered that the very circumscribed oolite deposites of northern Germany were only accidents in the great formation of muschelkalk.

The superposition of formations above each other being the fundamental base of sound geology, I shall commence by mentioning some localities where the muschelkalk rests on the new red sandstone. These citations might be very numerous, from the greater part of the large rivers of the N. and N.W. of Germany having hollowed their beds in the new red sandstone across the muschelkalk. It sometimes happens that this hollowing out takes place in a convexity of the inferior sandstone, as near Gottingen and Coburg. The surface of the new red sandstone is often uneven, and we then frequently see the muschelkalk mould its beds into inequalities, as near Detmold in Westphalia, near Stedfeld, near Eisenach, at Fachdorf, along the Werra, between Hoheneiche and Fatterode in Hesse; whilst elsewhere the horizontal limestone beds rest on a nearly even surface, as near Herrenhausen, near Pyrmont, between Memungen and Hilburghausen, and between Eislenben and Nordhausen; in France the band of muschelkalk which surrounds all the Vosges, with the exception of the northern part, reposes on the more or less irregular surfaces of the new red sandstone or variegated marls and gypsum, as near Bishmosheim, near Treves, &c.

The muschelkalk constitutes, in the north and west of Germany, a formation many hundred feet thick, whilst the zechstein or first secondary limestone never attains there more than a few fathoms in thickness.

This limestone is in rather thin beds, and very rarely affords sufficiently large blocks to be made use of as marble. Its horizontal or curved beds generally appear to contribute considerably to the form of the mountains composed of it; thus we sometimes see the rounded summits of this deposite or its platforms arise, the former from the convexities of the contorted beds (Stedtfeld, Detmold), and the latter from their horizontal position. These mountains sometimes present a steep slope, affording occasionally considerable escarpments by the side of the rivers, as near Fachdorf, in Memungen; these rocky escarpments are essentially distinguished from those of the oolitic heights, and an exact observer could not confound them with the cleft and indented precipices of the latter. It sometimes contains caverns, as in the Memungen, near Kloster-Fesser. This limestone is commonly compact and grey, and of a peculiar crystalline aspect, which is scarcely again found but in some beds of the oolite formation; it is occasionally rendered sublamellar by means of spathose organic remains.

The shells of this limestone are either petrified or in casts: their petrifaction is commonly calcareous and rarely siliceous, which is sometimes the case, as for example near Saarbruck. The shelly beds occur very generally, yet we may mention those of Mont Heimberg, near Gottingen, that of the environs of Walterhausen and Pyrmont; they contain innumerable remains of encrinites (En. vulgaris and liliformis), and a genus of zoophyte approaching the Isis.

The multilocular shells of this limestone are principally of the genera ammonites (am. nodosus, capricornus, dorsuosus, amalteus, &c. *Schlotheim*) and nautilus (N. bidarsatus, *Schlot.*) Among the bivalves we especially remark the mytili (M. eduliformis, socialis, incertus, costatus), chamæ and pectens. The terebratulæ (P. fragilis and communis) form true beds in it. The other bivalves of the Linnean genera Mya, Tellina, Donax, Venus, Arca, are much more rare.

Among the univalves, those of the genera buccinum, turbo, and trochus are the most common; yet these shells occur rather isolated; and we only here and there observe beds or masses of shells approaching the genus cerithium, as at the Budenberg, near Neustadt, in Hanover, &c.

Many other genera of univalves are also met with, though rarely, in this deposite; but it should be remarked that the greater part have only been found near Weymar, Phangelstad, Tonna, Jena, &c. i. e. in the midst of the great valley situated between the Hartz, the Thuringerwald, and the Erzgebirge.

It is there that M. Schotheim cites the dentalia (D. lœvis and torquatus), the helices of Linnæus (orientinus,) the neritæ (N. spirita and pagana), the patellæ (discoidea and mitrata); there also are found his lepas avirostris, his solen mytloïdes, his tellina anseps and minuta, his craniolites schroteri, and his curious bitubulites irregularis.

The mountain of Hecniberg, near Gottingen, and the environs of Hildesheim present many analogous rarities.

Remains of fishes have occasionally been observed in this limestone, especially the scales, and the remains of marine animals perhaps of genera approaching the manati; they are generally the remains of maxillary bones, the long bones of ribs, and vertebræ. I have observed organic remains bearing some resemblance in form to confervæ.

Fossils have also been remarked which rather belong to more recent strata, such as the asteria ophiurus. (Tentleben), which reappears in the quadersandstein, and the belemnites poxillosus, *Schl.* (Gottingen and Werkershausen) and the echinites pustulosus. (Eckorsleben). But it should be remarked that these fossils are found very rarely and in the very uppermost strata of the deposite, and that they should not be mentioned in a list of the characteristic fossils of the muschelkalk of Germany and France.

The encrinites, the terebratulæ, the ammonites, the remains of the Isis? some bivalves and univalves do not the less remain the characteristic and important fossils.

After this sketch of the general characters and fossils of the muschelkalk, we shall mention its principal varieties or subordinate beds.

The limestone sometimes passes into marl, especially in its upper strata and also decomposes into marl: it is occasionally very compact and nearly without shells, it is otherwise very shelly, so that it sometimes has the appearance of an aggregate of organic remains, and slightly resembles some beds of the oolite formation and zechstein; this occurs, for example, near Frankenhausen. A peculiar oolitic structure is met with in the lower strata of some localities, as at Bensdorf, Schorbe, near Ecinberg, and generally on the confines of Hanover and the small principalities of Westphalia, Buckeberg, &c.

Small granular masses are rarely found; beds are more frequently observed which are more or less cellular with angular cavities, nearly resembling those of some magnesian limestones of England; these limestones are yellowish, perhaps magnesian, and traversed by small calcareous veins (Pyrmont, Werkershauen, the Vosges).

Some beds of this limestone are blackish, brownish, coloured by hydrate of iron, and brownish red.

These varieties do not appear to be distributed in the deposite without a certain order; we always find the oolitic varieties in the lower parts, and especially in the localities where the marls of the new red sandstone do not alternate with a kind of oolite; above these limestones come the compact limestones with scattered fossils, and the beds of limestone filled with the remains of the Isis? then the limestones in which the terebratulæ especially abound, and which are sometimes slightly black, a variety which occurs with few shells among the lowest strata (the Buckeberg).

The yellowish cellular beds are among the upper parts and in the environs of Pyrmont, are covered by compact limestones, the upper strata of which contain small crystals of prismatic quartz. We also here and there observe small masses of sulphuret of lead, apparently rolled? in this limestone, sometimes accompanied by druses of crystallized quartz. (Heinberg, Pyrmont).

Small calcareous veins are often seen in the muschelkalk as also slight siliceous infiltrations, yet the latter are rare in Germany, and it is only among the lower strata of this limestone that we occasionally observe yellowish or greyish chert (silex corné), as at Hohenhagen, near Gottingen, at the Langenberg, near Coburg, and near Gotha. This circumstance also occurs on the western ridge of the Vosges chain, where this siliceous matter, more or less mixed with limestone, forms continuous beds, as near Bishmosheim.

The description we have just given applies to the band of muschelkalk which surrounds the Vosges, to the narrow chain of muschelkalk which extends from Warburg by Bielfeld into the Osnabruck, to the platform of muschelkalk to the north and west of the Hartz, to that of Hesse and of the great basin of Saxony and Thuringia, and to the great platform which extends from Hanau to near Stutgard, and which M. Keferstein has without any reason classed with the zechstein. On the eastern side of the great basin, of which the latter mass covers the western and northern edge, another limestone occurs which is equivalent to the muschelkalk, or in other words, the muschelkalk, on the western side of the Thuringerwald, here and there essentially changes its character; nearly in the same manner as the first secondary limestone (magnesian limestone) does on the sides of the same basin.

It is important to observe this singular change, since it may afford us the key by which to recognise this deposite in other countries; it more particularly takes place in the Coburg country. The muschelkalk forms a large platform on the north of this town, extending towards Meinungen and Neustadt; the Monts Langenberg, near Coburg, are still composed of the true muschelkalk, with siliceous nodules in its lower parts; but on the S.W. we observe only deposites of a very magnesian compact limestone of a whitish or whitish grey colour by the side of the oolite formation. It contains small veins of magnesian carbonate of lime, and very irregular nodules and strings of jasper, and a species of chert or even coarse chalcedony, of a red, grey, and white colour. The common fossils have altogether disappeared.

This limestone, which is connected, as we have said, with the new red sandstone by an arenaceous limestone, forms the top of Mont Eckersberg, of the Bucheberg, and of the monts on the east of the chateau of Coburg. It is seen near Bohrbach, Rogen, Lutzelbuch, and at Neuhofer-Muhle, from whence it extends towards Banz, occasionally covering the new red sandstone. On the east of Coburg, it forms a small platform from Oslau to the Mahnberg, it reappears near Ecinberg; more to the S.E. a very narrow crest of this limestone runs along small platforms of the true muschelkalk, and in the end, we see this dyke, some feet thick, abut against the latter limestone, near Kipfendorf, whilst the true muschelkalk, after having formed four small platforms between Coburg and Gestungshausen (the first on the E. of Rohsbach, the second on the E. of Kipfendorf, the third to the E. of Fechheim, and the fourth to the N. of Gestungshausen), also disappears and only shews itself again on the north in the environs of Baireuth.

In a word, we not only incontestably see there, the muschelkalk prolonged along the oolite chain of this part of Germany, but also the muschelkalk placed on the borders of a basin which has apparently had so much influence on the nature of this limestone, that as soon as we descend into it, we only see a magnesian limestone without fossils. The beds of new red sandstone, near Kipfendorf, at the foot of the Mahnberg, dip out of the basin to the S.E., and in the basin they become nearly vertical with a dip to the N.N.E. On the first parts of these beds rests a platform of true muschelkalk, whose breadth is narrowed to a few fathoms, from the 2 or 300 toises it had more north at Manchenroth; close to this, at only a few feet distance, a deposite of the magnesian limestone is found connected with the new red sandstone.

From the characters and anomalies of the deposite, it is possible that this limestone may be found in many localities where it has not yet been mentioned. It appears certain, that along the Bahmerwaldgebirge, it terminates at the platform of the environs of Baireuth, and that, on the other side of the basin, it is prolonged with interruptions along the oolite chain into Switzerland, where it still occupies some space near Basle, and there constitutes the rauchgrauer kalkstein of M. Merian.* It is afterwards united near Befort to the narrow band which it forms at the eastern foot of the Vosges, from thence to Alzey, whilst along the western side, from the combined observations of M. de Beaumont, M. Schmitz, and myself, this limestone extends from Lure to Vauvillers, Bourbon-les-Bains, Ligneville, Sarrebourg, and forms a great platform above the new red sandstone between Rosbach, Waldfishbach, and Forbach. It even rises higher north by Longeville and Trèves, and always occurs between the new red sandstone and the oolite formation.

* See Beytrage zur Geognosie, 1821.

In the other parts of France, I have only suspicions of its existence in the environs of Vitteaux, Rouvray, and Cussy-les-Forges, &c. Perhaps it also exists to the south of Nevers; the limestone near Aubenas, in the Vivarais, presents some of the characters, which distinguish it from the lias, which occurs more to the west of that town. It appears to be totally wanting in the west of France, as in England; but it occasionally appears united with the new red sandstone at the foot of the Pyrenees; thus in the department of the Landes, at the foot of Porci d'Arzet, in the commune of St. Pandelon, the upper part of the new red sandstone marls, with gypsum and saline springs, contains two beds of muschelkalk partly of a peculiar oolitic structure, and it is also seen above the same marls, on the foot of the mountains of the Conserans.

On the other hand, it is possible that this limestone even occurs in the Alps, particularly in those of Salzburg, for some limestones are known there, which are whitish or grey, granular or compact, with terebratulæ and pectens, and which from their position may one day be recognised as muschelkalk. On the southern side of the Alps, the Abbé Maraschini suspects its existence beneath the oolite formation of the Veronais; and in Hungary, some magnesian limestones, like those of the environs of Bude, may, from their position and nature, be sooner or later referred to it.

Third Secondary Sandstone or Quadersandstein.

The second secondary limestone is in Germany covered by an arenaceous deposite, named quadersandstein by the Germans. This deposite is as little known to foreign Geologists as that of the muschelkalk, they have not yet learned to assign it its proper place among the secondary formations, and they have either not recognised it in their own countries, or have compared it to deposites altogether different, as for example, the sandstone of Fontainebleau.

The obscurity that still envelopes this formation depends

on its great rarity in other countries, and this again shews us that in geology, in order to recognise a formation in any country, the deposite should have been studied in the country where it is most completely developed, this principle is especially applicable to the recent secondary formations, which are more than others the deposites of great basins or sinuosities more or less separated from each other.

Thus if it is not surprising that a French geologist, who has not studied the quadersandstein in Germany, should not recognise it in France where it is scarce; on the other hand, we should not be surprised that a German, who has not visited the well characterized tertiary rocks in the north of France, should find himself embarrassed how to class certain scattered deposites in his own country. In Europe, for the same reasons, the muschelkalk and zechstein formations should be studied in Germany, whilst the oolite formation and chalk should be seen in Switzerland, France, and England, and not in the north of Germany; the trachytic deposites in Hungary and not on the banks of the Rhine; the extinct volcanoes and patches of ancient basaltic *coulées* in Auvergne and in the Vivarais and not at the Eiffel, Eger, or in the Mittelgebirge; the basaltic Huttonian cones in Hesse and Thuringia and not in Bohemia; the trap rocks of the red sandstone in Scotland and the Palatinate, and not at Noyant or Figeac in France, &c.

The quadersandstein or third arenaceous secondary formation rests on the true muschelkalk, between Hildesheim and Dickholzen, near Helmstadt, at the chapel of Lindach, between Wipfeld and Lindach, not far from Schevenfurt on the Mein, at Stegerwald, near Hassfurt, and muschelkalk has been attained in a well in the quadersandstein of the garden of Nesselhof, near Gotha. We sometimes see, in the environs of Coburg, a magnesian variety of muschelkalk dip beneath the quadersandstein, as near Oferfullbach. In Westphalia, we see the beds of the muschelkalk band, extending from Steinheim, by Bielfeld, to Hilter, dip to the N. and the oolite marls alternating with the quadersandstein rest on it, possessing the same dip. Near Pyrmont, we even see these marly and arenaceous alternations united to the muschelkalk by alternating with two or three beds of a grey compact limestone, identical with some muschelkalks, but without any organic remains. A similar alternation is seen at the foot of the Bierberg, near Lude, to the south of Pyrmont; the limestone is there accompanied by small portions of marl highly impregnated with yellow hydrate of iron. It naturally happens that the quadersandstein occasionally rests on the new red sandstone; this occurs for example, near Opferbaum, between Scheveinfort and Wurtzburg, where it appears to rest immediately on the gypsum of the new red sandstone marls. The same thing happens on the north of the Hartz, and particularly along the western side of the oolite chain of northern Bavaria, where we pass immediately from the new red sandstone to the quadersandstein, as for example, on the south of Nurnberg. Yet it is much more easy to distinguish them than it is to separate the red sandstone from the new red sandstone deprived of marls.

In Bohemia, where the quadersandstein often covers the coal measures or red sandstone, for example near Brandies, &c. in the Erzgebirge (Gruntenburg, Nieder, Schona) and on the banks of the Elbe, on the limits of Austria and Saxony, where it rests on gneiss, it cannot be confounded with any other formation.

The quadersandstein may be described as a generally fine sandstone,* composed of small rounded grains of quartz, and occasionally mixed with scales of silvery mica, which are sometimes distributed in parallel and interrupted laminæ. The cement of this sandstone is argillaceous or argillo-ferruginous; it is generally more slightly aggregated than other sandstones, and this sandstone also much resembles modern alluvions; yet, in many cases, it has been hardened, like the tertiary sandstones, by a calcareous cement, or more rarely by silex.

In the first case, this rock surprisingly resembles the

* See the exact descriptions given by M. Haussman, Norddeutsche Beitrage, &c. p. 68. and Driburger Taschenbuch, 1816.

upper marine sandstone of Fontainebleau, as at the Schevelbenwald, and at the top of the Kotersburg, near Pyrmont; we also sometimes see, as at Fontainebleau, crystals of *(inverse)* carbonate of lime, as for example, at Blankenburg. The quartz occasionally forms small veins.

It very often happens that parts of this sandstone are decomposed into a yellow or white sand, as at the foot of Mont Bomberg at Pyrmont, and at the Kontersberg, where this sand resembles the iron sand of the English. These sands, in other localities, produce an extensive moveable surface, as between Blankenburg and Halberstadt, and especially in the N.E. of Bohemia.

The colours of the quadersandstein are white, whitish yellow, yellow, brownish, and rarely of a roseate tint; the first varieties abound in the north of Bohemia, Saxony, and on the north of the Hartz, while the yellow and brownish varieties are met with round the oolite chain of the S.W. of Germany; the latter do not furnish as good building stones as the others.

The subordinate beds of this formation are not numerous; in the lower strata we often observe coarse beds, in which quartz pebbles are found associated with pieces of flinty slate and Lydian stone; this is seen near Vigy not far from Metz, and in the Erzgebirge, for example at Kisibel, &c. It happens in the latter chain, near Freyberg, that these beds contain numerous pieces of white granular quartz, identical with the quartzose gangues of many metalliferous veins in the gneiss.

Sometimes we also observe slightly marly beds in the quadersandstein; this is seen in the sandstone of Pirna, Gotha, and Silesia. We often observe in the upper beds a thickness of some feet or fathoms occupied by alternations of ferruginous yellow sandstone with clays and clayey marls, which are grey, blueish grey, greenish grey, and even reddish, as near Oberfulbach in Coburg, and at Vigy near Metz. The beds of clay are sometimes advantageously employed for pottery, as at Kipfendorf (Coburg), but these beds are rarely thick enough to be worked in the same man ner as those of the plastic clay. Yet it sometimes happens that we may be embarrassed in deciding whether a deposite belongs to one or the other of these formations, especially when the marls of the quadersandstein present the undulations of the tertiary beds, when they are on the surface, or when they are in the vicinity of true plastic clays.

These argillaceous beds moreover sometimes contain small masses of lignites, for vegetable remains are not foreign to the quadersandstein, they are on the contrary very characteristic of it, serving to distinguish it from the other two more ancient secondary sandstones.

These vegetable remains are wood and monocotyledonous plants; * the first are changed into sandstone or infiltrated by silex, or else they present varieties of mineral carbon and lignite. These vegetable remains especially abound in certain beds, and give a peculiar aspect to these sandstones, as is seen at Kipfendorf and Blumenroth in Coburg, and at Vigy near Metz. Sometimes these sandstones have a grey tint, or contain small strings of lignites, as near Quedlinburg and Pirna.

Beds of lignite are even worked in this sandstone, as for example, in Coburg, to the E. of Spittelstein, and along the western side of the oolite chain in the S.W. of Germany.

These deposites of lignites are most abundant in Westphalia, and they have been profitably worked in the Buckeburg. In this part of Germany, as we shall see below, the quadersandstein is united with the lower marls of the oolite formation; we observe a great part of the space between Osnabruck, Bielfeld, Vlotho, and Buckeburg, occupied by alternations of marls and sandstones. These sandstones are sometimes identical with those of the quadersandstein, as for example, the coarse bed traversed by strings of fibrous carbonate of lime, which is worked at the Porta Westphalica (a defile of the Weser, near Minden) and the sandstone of Hall, in Bielfeld.

* It will be observed that M. von Humboldt says these plants are dicotyledons. (Trans.)

Yet the greater part are more or less compact sandstones, coloured grey, violet grey, greenish grey, and brownish, hardened by marls or by ferruginous infiltrations, as is seen near Herford, and the patches of this formation which occur on the S. and W. of Pyrmont, near Luntorf, Rudsick, Falkenhagen, &c.; some certain small masses of marl, and at a distance resemble coal grit.

These lignite sandstones, containing pyrites and alternating with slate clays and marls, which are shelly in their upper strata, cannot be confounded with the coal measures; in the first place, because the combustible is always only a bituminous wood, a pyritous mineral carbon, or else a jet, which rarely seems to pass into certain varieties of pitch coal (Minden Buckeburg.)*

The abundance of marine fossils, of marls and slate clays, the nature of the sandstone, the small number of coal beds, and the nearly total want of those disruptions of beds which is observed in the true coal measures, are sufficient characters to distinguish this formation from any other.

The alternation of this sandstone with the lower strata of the oolite formation is not confined to Westphalia, for traces of it are seen in Coburg, where, at Blumenroth, we observe two beds of the same grey compact sandstone, alternating with marls, between the true quadersandstein and the lower and shelly part of the oolite formation. Further to the south, similar facts occur more distinctly beneath the lias of southern Bavaria and Wurtemberg. Nests of iron pyrites are occasionally found, as near Bohrbach in Coburg.

The fossils of the quadersandstein are abundant; we have already mentioned siliceous or bituminous wood, and impressions of monocotyledonous plants †; the siliceous wood is particularly abundant in Coburg, and is sometimes coloured green by nickel. The impressions of wood and pieces of plants are frequent; sometimes the vegetables have disappeared and left no void spaces (Gittersen, Coburg); their position is either horizontal or inclined.

* See Wurzer, Analyse der Schwefelquellen zu Nendorf, 1815.

† Dicotyledons, Humboldt, (Trans).

These impressions have never presented me the singular figures which characterize the plants of the coal measures; on the contrary, both woods and plants appeared to me much more analogous to the actual European vegetation than to the plants of the coal measures.

Well preserved impressions are very rare, caused probably by the want of the fine coal measure clays, and by the different origin of the deposite; yet impresssions of plants may be observed which bear some resemblance to reeds (Luntorf, near Pyrmont), very distinct leaves resembling at a distance those of the hazel and walnut tree, as at Mont Heidelberg near Blankenberg, and at Wulfenbuttel. I have even seen in the superb cabinet of Baron von Schlotheim, vegetable impressions resembling palms (palmacites annulatus, canaliculatus, and absoletus, &c. *Schl.*) and even plants resembling ferns, or of the division lycopodiolithes of M. *Schlotheim* (L. cœspitosus) (Gotha). The same naturalist mentions carpolithes. Marine organic remains are tolerably abundant, especially in some localities and certain beds; but they are in general only casts, and are rarely siliceous or chalcedonic petrifactions, as at the Platenberg near Blankenburg.

The most frequent fossils of this kind would appear to be pectens, (p. punctatus, radiatus, longicollis, anomalus, *Schlot*:); they are particularly found in some beds of the quadersandstein of Silesia and Pirna. In the last locality we also see shells of the genus Venus, oysters (ostrea labiata *Knorr*) and mytili.

On the north of the Hartz, the turbinites obvolutus, *Schl*: and regensbergensis, *Knorr*, have been long known near Blankenberg, as also in the Halberstadt, where it is associated though rarely with ostræa crista galli.

Volutæ and bullæ are mentioned as occurring in the environs of Halberstadt, myæ (m. musculoides, *Schl*:) exist in the Seeberg, at Gotha. The rare asteria lumbricalis *Schl*: occurs in nearly the upper strata of the Coburg quadersandstone (Gossenburg); this bed also contains indeterminate bivalves.

The sandstone of Hildesheim sometimes contains terebratulæ (t. acuta *Schl*:), and between Stoffenheim and Teilhosen in Bavaria, I have seen, in the coarse ferruginous sandstone, immediately beneath the lias, a great abundance of the gryphites arcuata and belemnites.

Echinites and pinnites (p. diluvianus) are mentioned as rarities from Pirna, as also the remains of encrinites; yet the latter are sometimes very abundant in some of the upper beds, and are mixed with bivalves (mytilus?) as in the Staffelsberg, near Staffelstein, and at Blumenroth in Coburg.

Certain singular prominencies of this sandstone might lead us to suspect the existence of crustaceæ.

This formation is not only distinguished by the nature of its rocks and fossils, but it very often constitutes mountains of a very peculiar form, in consequence of its irregular decomposition or cementation. Thus when this formation is not covered by the oolite formation, as in Bavaria, or else by chalk, it presents a suite of indented crests, with very singular rounded sections; such is for example the Devil's Wall or the Taufelmauer, between Blankenburg and Halberstadt, and the indented heights along the Elbe, between Pirna and Petchen.

When this formation has been much destroyed and a few patches of it only remain, we then observe it to form simple walls, as at Goslar, or singularly shaped blocks, as at Hackstein, near Hirchberg, in Bohemia, &c.

Its valleys are deep, with very steep and nearly vertical slopes, at least when the deposite has been considerable, as between Tanneberg and Bohmish, Kamnitz, and near Oschitz in Bohemia, or else the valleys are very much hollowed out, the bottoms being occupied by more ancient formations, whose heights are alone covered by sandstone, as in many localities in the N.E. of Bohemia, and northern Bavaria. After having described this formation as exactly as I am able, I may be permitted to pass in review its known localities.

This formation had long been described in Bohemia, and it was there known to cover the coal measures and red sandstone, and to be occasionally hidden beneath some fathoms of green sand or chloritous chalk or planerkalk. It begins to appear, in that country, en the frontiers of Silesia, at the Heuscheur, and at an elevation of 2893 feet above the level of the sea. It extends from thence throughout the north of Bohemia, and its southern limit should not certainly be placed further back than Eypel, Arnau, and Jung Bunzlaumelnik. It still occasionally appears on this side the Elbe, as near Raudnitz, Prague, Bandeis, &c.: some patches are observed further south, in the sands of Konigingratz, and even perhaps in Moravia.

After having been covered by basalts, and confined between the basalts of the eastern part of the Mittelgebirge, and the primitive rocks and red sandstone of the Riesengebirge, the quadersandstein extends on both sides of the Elbe towards Pirna, ascending along some valleys of the Erzgebirge, where patches of it are sometimes seen. It extends on the other side into Lusace, and is seen on this side of the granitic mass near Ullersdorf, and in numerous localities in Silesia.

In the great valley between the Erzgebirge, the Thuringerwald and the Hartz, it is only known near Gotha to the north of the Seeberg, at Boxstedt, and near Walterhausen, whilst on the north of the Hartz, it abounds near Helmstadt, and forms a nearly continuous wall from Quedlinburg to Wernigerod; it is only seen in isolated mounts beyond that to Hildesheim, being much covered by cretaceous deposites, as at Gorlar, Salzbetfurth, and Hildesheim.

It is found in patches in the upper part of the valley of the Leine, as for example, near Guttersen, &c. it is only observed in Westphalia, where we have already mentioned the great space that it occupies with the oolite marls. It is seen no more after that on the west, except near Aix la Chapelle, where perhaps a patch exists tolerably well characterized as to the rock and its vegetable remains.

It begins to shew itself, in the S.W. of Germany, occasionally along the Mein, near Schweinfurt, and on the south of Coburg; it then extends along both sides of the oolite chain of that country; it goes on one side towards Amberg, and on the other into Wurtemberg, where it alternates with lias. This band is more or less broad, according as the sandstone is more or less covered by the oolite formation, or more or less preserved above the new red sandstone; it is of considerable extent between Roth, Weissenburg, and Nuremberg.

I am only acquainted with it at present in the eastern and southern parts of France; in the first place it exists in great force, according to M. von Buch, to the north of Luxembourg, between that town, Feltz, and Alfdorf; there are three small deposites of it to the E. of Metz, one between Vigy and St. Hubert, a second on the north of Bertoncourt, and a third to the south of Ketange. It should be remarked that all these portions are situated in the ramifications of the great valley of the Moselle, on the western side of which we also occasionally find the ferruginous varieties, which are even worked as hydrate of iron ore, as for example, at Hayonge. Further south, I am only acquainted with some traces of it near Vic and in the environs of Vitteaux. Some fragments have been found by M. de la Jonquere, in the midst of the lias of Mezieres, and other shelly varieties have been discovered by M. de Beaumont, with the same rock, not far from Buxweiler.

In the Jura, the marls which are slightly sandy replace the quadersandstein, and alternate with the inferior oolite marls.

In the Pyrenees, the quadersandstein forms very considerable strata beneath the whole band of the oolite formation which runs along their northern foot; it always presents the same quartzose sandstones, which are more or less marly, micaceous, or slaty. They contain marine organic remains, and fossil vegetables; they alternate with the lower part of the oolite formation, and contain traces of lignites like the latter limestones. I shall content myself by mentioning as examples, the quadersandsteins on the north of Navarreins, and these of St. Paul and Nalzen, in the Pays de Foix.

This formation does not appear to exist along the northern side of the Alps, except in the Alps of Swabia, and perhaps Bavaria; it is certain that the sandstones of the environs of St. Galles with the gryphites spiratus and many other peculiar shells, have greatly the appearance of belonging to this formation, that is, if they do not form part of the green sand which is the only sandstone that may sometimes be confounded with the quadersandstein. On the other hand, the connection we have seen to exist between the lower oolite marls and the sometimes carbonaceous sandstones of our deposite, leads us to suppose that many deposites of carbonaceous matter or lignites, especially with marine remains, noticed on the foot of the Alps of Swabia, may belong to this kind of formation. We may one day unite with it the masses of carbonaceous matter which occur along the Alps, between France and Piedmont, as the mass of Entrevener, those on the environs of Grenoble, &c.

FINIS.

LONDON: PRINTED BY WILLIAM PHILLIPS,
GEORGE YARD, LOMBARD STREET.

A SYNOP

The Names having the letter B

ENGLISH.	FRENCH.	GERMAN.
ALLUVIUM & DILUVIUM...	TERRAIN DE TRANSPORT & d'ALLUVION (B). *Terrain de Transport* (Daubuisson.)	AUFGESCHWEMMTE-GEBIRG
Superior Order (Conybeare)..... *Tertiary Rocks*........	Terrain de Sediment Superieur (B)... *Terrain Tertiaires*.........	Jungstes Kalkstein Gebilde & Braunkohlen Formation (Kefer-stein).
Upper fresh-water formation	Troisième terrain d'eau douce. (B.)..	Süsswasser-Formation (Keferstein
Upper marine formation.........	Deuxième terrain marin. (B.) .....,	
Lower fresh-water formation	Deuxième terrain d'eau douce. (B.).	Knochenführender Gyps. &c. (Ke
Lower marine formation.........	Premier terrain marin. (B.)........	Ceritenkalkstein. (Keferstein.)
London Clay................	*Calcaire grossier.*	
Plastic Clay..................	Premier terrain d'eau douce. (B.)... *Argile Plastique.*	Plastischer Thon (Keferstein)
SUPERMEDIAL ORDER. (C.) *Secondary Rocks*	TERRAINS SECONDAIRES....	FLÖTZGEBIRGE.
Chalk..........................	Craie..........................	Kreide-Formation.
a Upper or flinty chalk.....} *b* Lower chalk ....}	*a.* Craie blanche ou Superièure ...	
c Grey chalk & chalk marl....	*b.* Craie tufau	
Green Sand ....} Iron Sand.................}	*c.* Craie inferièure .... *Craie Chloritée.* *Glauconie crayeuse.* (B) *Grès & Sables verts & ferrug.*	
Oolite formation	Calcaire du Jura	Jurakalk formation.
Lias....................	Calcaire à Gryphites.	
	Quadersandstein	Quadersandstein.
	Muschelkalk	Muschelkalk.
Red Marl & new Red or Salife-rous sandstone............}	Grès bigarré	Bunter-sandstein.
Magnesian limestone	Calcaire Alpin................. *Zechstein.* (Humboldt)	Alpenkalkstein. *a.* Zechstein, &c.
New Red Conglomerate	Pséfite rougeatre. (B)........... *Grès rouge.*	Rothe-todte-liegende.
New Red Porphyry	Porphyre du Grès Rouge......... *Porphyre Secondaire.*	Porphyr gebirge. (Keferstein).
MEDIAL or CARBONIFEROUS ORDER (Conybeare).	Generally considered by Foreign Geologists as referable partly to the p[re]ding, partly to the following class.	
Coal Measures...............	Terrain Houiller .... *Grès rouge.* (Humboldt).	Steinkohlengebirge.
a. Slate clay	*a.* argile schisteuse	*a.* Schieferthon.
b. Coal Grit................	*b.* grès des Houillères......... *Psammite.* (B).	*b.* Kohlensandstein.
Mountain or Carboniferous Limest.	Calcaire de Trans. (O. d'Halloy, &c.)	
Old Red Sandstone............	Grauwacke (Humboldt, &c.)	Grauwacke.

, TABLE OF EQUIVALENT FORMATIONS, BY H. T. DE LA BECHE, Esq

are adopted by Brongniart in the succeeding "Table:"—Such as are printed in *Italics*, are the Synonyms o

ENGLISH.	FRENCH.	GERMAN.
'RIMORDIAL ROCKS.	TERRRAINS PRIMORDIAUX. (Omalius d'Halloy.)	
'ary Rocks (Macculloch.) *edial & Inferior Orders.* onybeare) .... *sition & Primitive Rocks.*	*Terrains intermédiaires & primitifs* *Terrains de Transition & D°*	Übergangsgebirge & Urgebirge.
wacke	Grauwacke .... *Traumate* (Daubuisson) .... *Psammite* (B)	Grauwacke.
vacke Slate..............	Grauwacke schistoïde............. *Schiste Traumatique.* (Daubuisson)	Grauwackenschiefer.
ition Limestone *medial Limestone*	Calcaire de Transition .... *Calcaire intermédiare.*	Übergangskalkstein.
Slate	Schiste Alumineux................. *Ampélite Alumineux.*	Alaunschiefer.
stone Slate	Schiste coticule .... *Schiste novaculaire* *Schiste à aiguiser.*	Wetzschiefer.
Slate....................	Schiste Silicieux .... *Jaspe Schistoïde* (B)	Kieselchiefer.
tine	Serpentine .... *Ophiolite.* (B)	Serpentin.
ge Rock..................	Euphotide. (Haüy)...............	Schillerfels. *Gabbro.* (Von Buch).
stone	Diabase (B)	Grünstein.
stone Slate	Diabase schistoïde. (B.).........	Grünsteinschiefer.
z Rock	Roche de Quarz .... *Quartzite* (B. & Bonnard)	Quarzfels.
Slate	Schiste Argileux .... *Phyllade* (Daubuisson.)	Thonschiefer.
te Slate	Schiste chloriteux	Chloritschiefer.
se Slate	Schiste talqueux.................	Talkschiefer.
ist	Steaschiste (B)	
lende Rock...............	Amphibolite (Daubuisson)	Hornblendegestein.
ende Slate	Amphibolite schisteuse	Horblendschiefer.
ive Limestone............ *nular Limestone*.........	Calcaire primitif .... *Calcaire grenu.*	Urkalkstein.
Slate .... *aceous Schist* (Macculloch) .	Schiste micacé .... *Micaschiste* (B).	Glimmerschiefer.
act & Granular Felspar meson)	Eurite. (Daubuisson)	Weisstein.
stone (Jameson)		
	Leptenite. (Haüy)	Hornfels.
s	Gneis	Gneuss.
te	Granite	Granit.
. Graphic Granite	*var.* Pegmatite. (Haüy).	
. Protogine	*var.* Protogine. (Jurine).	

:AL TABLE OF EQUIVALENT FORMATIONS, BY H. T. DE LA BECHE, F

xed are adopted by BRONGNIART in the succeeding "Table:"—Such as are printed in *Italics*, are the Synonyr

ENGLISH.	FRENCH.	GERMAN.
PRIMORDIAL ROCKS.	TERRRAINS PRIMORDIAUX. (Omalius d'Halloy.)	
Primary Rocks (Macculloch.) *Submedial & Inferior Orders.* (Conybeare) .... *Transition & Primitive Rocks.*	*Terrains intermédiaires & primitifs* *Terrains de Transition & D°*	Übergangsgebirge & Urgebirge.
Greywacke	Grauwacke .... *Traumate* (Daubuisson) .... *Psammite* (B)	Grauwacke.
Greywacke Slate...............	Grauwacke schistoïde.............. *Schiste Traumatique.* (Daubuisson)	Grauwackenschiefer.
Transition Limestone *Submedial Limestone*	Calcaire de Transition .... *Calcaire intermédiare.*	Übergangskalkstein.
Alum Slate	Schiste Alumineux................ *Ampélite Alumineux.*	Alaunschiefer.
Whetstone Slate	Schiste coticule .... *Schiste novaculaire* *Schiste à aiguiser.*	Wetzschiefer.
Flinty Slate..................	Schiste Silicieux .... *Jaspe Schistoïde* (B)	Kieselchiefer.
Serpentine	Serpentine .... *Ophiolite.* (B)	Serpentin.
Diallage Rock.................	Euphotide. (Haüy)................	Schillerfels. *Gabbro.* (Von Buch).
Greenstone	Diabase (B)	Grünstein.
Greenstone Slate	Diabase schistoïde. (B.)..........	Grünsteinschiefer.
Quartz Rock	Roche de Quarz .... *Quartzite* (B. & Bonnard)	Quarzfels.
Clay Slate	Schiste Argileux .... *Phyllade* (Daubuisson.)	Thonschiefer.
Chlorite Slate	Schiste chloriteux	Chloritschiefer.
Talcose Slate	Schiste talqueux..................	Talkschiefer.
Steachist	Steaschiste (B)	
Hornblende Rock...............	Amphibolite (Daubuisson)	Hornblendegestein.
Horblende Slate	Amphibolite schisteuse	Horblendschiefer.
Primitive Limestone........... *Granular Limestone*..........	Calcaire primitif .... *Calcaire grenu.*	Urkalkstein.
Mica Slate .... *Micaceous Schist* (Macculloch) .	Schiste micacé .... *Micaschiste* (B).	Glimmerschiefer.
Compact & Granular Felspar (Jameson) .... Whitestone (Jameson)	Eurite. (Daubuisson)	Weisstein.
	Leptenite. (Haüy)	Hornfels.
Gneiss	Gneis	Gneuss.
Granite	Granite	Granit.
var. Graphic Granite	*var.* Pegmatite. (Haüy).	
var. Protogine	*var.* Protogine. (Jurine).	

F.R.S. &c.

ose immediately preceding them.

ENGLISH.	FRENCH.	GERMAN.
RAP ROCKS	...ROCHES TRAPPÉENNES....	...TRAPP-GEBIRGSARTEN.
erlying Rocks (Macculloch).		
cke	Wacke	Wacke.
ystone........................	Argilolite (B)	Thonstein.
kstone	Phonolite. (Daubuisson)...........	Klingstein.
npact Felspar	Feldspath Compacte .... *Petrosilex*. (B).	Dichter Feldspath.
hstone	Rétinite. (B.)	Pechstein.
nblende Rock	Amphibolite......................	Hornblendegestein.
ilt	Basalte	Basalt.
erite	Dolérite..........................	Dolerit.
ugit Rock (Macculloch).		
ygdaloïd	Amygdaloïde	Mandelstein.
r. of Amygdaloid	Variolite	Blatterstein.
ean	Cornéenne *Aphanite*.	
enstone	Diabase (B).......................	Grünstein.
ite	Syénite	Sienit.
Porphyry	Porphyre	Porphyr.
Clay Porphyry	Porphyre terreux	Thonporphyr.
Claystone Porphyry	*Argilophyre* (B)	*Thonstein porphyr*.
Clinkstone Porphyry	Phonolite.........................	Klingstein porphyr.
Felspar Porphyry	Porphyre euritique .... *Porphyre* (B).	Feldspath porphyr.
Pitchstone Porphyry.........	Stigmite (B)	Pechstein porphyr.
	Mélaphyre (B) .... *Porphyre noir*.	Trapporphyr.
Porphyritic Greenstone ..... *Greenstone Porphyry*.	Diabase porphyroïde............	Grünstein-porphyr
Trap-Tuff	....	Trap-tuff.
VOLCANIC ROCKS......	...TERRAINS VOLCANIQUES..	.. VULCANISCHE-GEBIRGE.
hyte	Trachyte	Trachit.
lstone	Perlite	Perlstein.
lt..............................	Basalte	Basalt.
....	Lave..............................	Lava.
Compact lava	Lave compacte.	
Scoriform lava	Lave scoriacée (B).	
Porphyritic lava	Lave porphyroïde (B).	
anic Amygdaloid...........		
dian	Obsidienne........................	Obsidian.
Obsidian Porphyry	Stigmite (B)	*a*. Obsidian porphyr.
ice	Ponce..............................	Bimstein.
anic Conglomerate	Brèche Volcanique	Vulcanischen Breccien.
anic Tuffa	Tuf Volcanique	Vulcanische Tuff.

Printed in the United States
By Bookmasters